Springer Undergraduate Mathematics

D0071139

Springer

London
Berlin
Heidelberg
New York
Barcelona
Budapest
Hong Kong
Milan
Paris
Santa Clara
Singapore
Tokyo

Advisory Board

Other books in this series

T.S. Blyth and E.F. Robertson

Basic Linear Algebra

 Springer

Professor Thomas S. Blyth
Professor Edmund F. Robertson

School of Mathematics and Computer Science, University of St Andrews,
North Haugh, St Andrews KY16 9SS, UK

Cover illustration elements reproduced by kind permission of:

Aptech Systems, Inc., Publishers of the GAUSS Mathematical and Statistical System, 23804 S.E. Kent-Kangley Road, Maple Valley, WA 98038, USA. Tel: (206) 432 - 7855 Fax (206) 432 - 7832 email: info@aptech.com URL: www.aptech.com

American Statistical Association: Chance Vol 8 No 1, 1995 article by KS and KW Heiner 'Tree Rings of the Northern Shawangunks' page 32 fig 2

Springer-Verlag: Mathematica in Education and Research Vol 4 Issue 3 1995 article by Roman E Maeder, Beatrice Amrhein and Oliver Gloor 'Illustrated Mathematics: Visualization of Mathematical Objects' page 9 fig 11, originally published as a CD ROM 'Illustrated Mathematics' by TELOS: ISBN 0-387-14222-3, german edition by Birkhauser: ISBN 3-7643-5100-4.

Mathematica in Education and Research Vol 4 Issue 3 1995 article by Richard J Gaylord and Kazume Nishidate 'Traffic Engineering with Cellular Automata' page 35 fig 2. Mathematica in Education and Research Vol 5 Issue 2 1996 article by Michael Trott 'The Implicitization of a Trefoil Knot' page 14.

Mathematica in Education and Research Vol 5 Issue 2 1996 article by Lee de Cola 'Coins, Trees, Bars and Bells: Simulation of the Binomial Process page 19 fig 3. Mathematica in Education and Research Vol 5 Issue 2 1996 article by Richard Gaylord and Kazume Nishidate 'Contagious Spreading' page 33 fig 1. Mathematica in Education and Research Vol 5 Issue 2 1996 article by Joe Buhler and Stan Wagon 'Secrets of the Madelung Constant' page 50 fig 1.

ISBN 3-540-76122-5 Springer-Verlag Berlin Heidelberg New York

British Library Cataloguing in Publication Data
Blyth, T.S. (Thomas Scott)
   Basic linear algebra. - (Springer undergraduate mathematics
   series)
   1. Algebra, Linear
   I. Title   II. Robertson, Edmund Frederick
   512.5
ISBN 3540761225

Library of Congress Cataloging-in-Publication Data
Blyth, T.S. (Thomas Scott)
   Basic linear algebra / T.S. Blyth and E.F. Robertson.
      p.   cm. -- (Springer undergraduate mathematics series)
   Includes index.
   ISBN 3-540-76122-5 (Berlin : pbk. : acid-free paper)
   1. Algebra, Linear. I. Robertson, E.F. II. Title. III. Series.
QA184.B597 1998   97-28661
512'.5--dc21   CIP

Printed in Great Britain
2nd printing, 1998
3rd printing, 2000
4th printing 2001

Typesetting: Camera ready by author (TSB)
Printed and bound at the Athenæum Press Ltd., Gateshead, Tyne & Wear
12/3830-543 Printed on acid-free paper SPIN 10832001

Preface

The word 'basic' in the title of this text could be substituted by 'elementary' or by 'an introduction to'; such are the contents. We have chosen the word 'basic' in order to emphasise our objective, which is to provide in a reasonably compact and readable form a rigorous first course that covers all of the material on linear algebra to which every student of mathematics should be exposed at an early stage.

By developing the algebra of matrices before proceeding to the abstract notion of a vector space, we present the pedagogical progression as a smooth transition from the computational to the general, from the concrete to the abstract. In so doing we have included more than 125 illustrative and worked examples, these being presented immediately following definitions and new results. We have also included more than 250 exercises. In order to consolidate the student's understanding, many of these appear strategically placed throughout the text. They are ideal for self-tutorial purposes. Supplementary exercises are grouped at the end of each chapter. Many of these are 'cumulative' in the sense that they require a knowledge of material covered in previous chapters. Solutions to the exercises are provided at the conclusion of the text.

We are greatly indebted to Dr Jie Fang for assistance with the proofreading and for checking the exercises.

The present text is a greatly enhanced version of the now unavailable volume 2 in the series *Essential Student Algebra* published by Chapman & Hall in 1986. We acknowledge the privilege of having it included in the launch of the Springer SUMS series.

T.S.B., E.F.R.
St Andrews

Foreword

The early development of matrices on the one hand, and linear spaces on the other, was occasioned by the need to solve specific problems, not only in mathematics but also in other branches of science. It is fair to say that the first known example of *matrix methods* is in the text *Nine Chapters of the Mathematical Art* written during the Han Dynasty. Here the following problem is considered:

There are three types of corn, of which three bundles of the first, two bundles of the second, and one of the third make 39 *measures. Two of the first, three of the second, and one of the third make* 34 *measures. And one of the first, two of the second, and three of the third make* 26 *measures. How many measures of corn are contained in one bundle of each type?*

In considering this problem the author, writing in 200BC, does something that is quite remarkable. He sets up the coefficients of the system of three linear equations in three unknowns as a table on a 'counting board'

$$
\begin{array}{ccc}
1 & 2 & 3 \\
2 & 3 & 2 \\
3 & 1 & 1 \\
26 & 34 & 39
\end{array}
$$

and instructs the reader to multiply the middle column by 3 and subtract the right column *as many times as possible*. The same instruction applied in respect of the first column gives

$$
\begin{array}{ccc}
0 & 0 & 3 \\
4 & 5 & 2 \\
8 & 1 & 1 \\
39 & 24 & 39
\end{array}
$$

Next, the leftmost column is multiplied by 5 and the middle column subtracted from it *as many times as possible*, giving

$$
\begin{array}{ccc}
0 & 0 & 3 \\
0 & 5 & 2 \\
36 & 1 & 1 \\
99 & 24 & 39
\end{array}
$$

from which the solution can now be found for the third type of corn, then for the second and finally the first by back substitution. This method, now sometimes known as *gaussian elimination*, would not become well-known until the 19th Century.

The idea of a *determinant* first appeared in Japan in 1683 when Seki published his *Method of solving the dissimulated problems* which contains matrix methods written as tables like the Chinese method described above. Using his 'determinants' (he had no word for them), Seki was able to compute the determinants of 5×5 matrices and apply his techniques to the solution of equations. Somewhat remarkably, also in 1683, Leibniz explained in a letter to de l'Hôpital that the system of equations

$$
\begin{aligned}
10 + 11x + 12y &= 0 \\
20 + 21x + 22y &= 0 \\
30 + 31x + 32y &= 0
\end{aligned}
$$

has a solution if

$$
10 . 21 . 32 + 11 . 22 . 30 + 12 . 20 . 31 = 10 . 22 . 31 + 11 . 20 . 32 + 12 . 21 . 30 .
$$

Bearing in mind that Leibniz was not using numerical coefficients but rather

two characters, the first marking in which equation it occurs, the second marking which letter it belongs to

we see that the above condition is precisely the condition that the coefficient matrix has determinant 0. Nowadays we might write, for example, a_{21} for 21 in the above.

The concept of a *vector* can be traced to the beginning of the 19th Century in the work of Bolzano. In 1804 he published *Betrachtungen über einige Gegenstände der Elementargeometrie* in which he considers points, lines and planes as undefined objects and introduces operations on them. This was an important step in the axiomatisation of geometry and an early move towards the necessary abstraction required for the later development of the concept of a linear space. The first axiomatic definition of a linear space was provided by Peano in 1888 when he published *Calcolo geometrico secondo l'Ausdehnungslehre de H. Grassmann preceduto dalle operazioni della logica deduttiva*. Peano credits the work of Leibniz, Möbius, Grassmann and Hamilton as having provided him with the ideas which led to his formal calculus. In this remarkable book, Peano introduces what subsequently took a long time to become standard notation for basic set theory.

Peano's axioms for a linear space are

1. $a = b$ *if and only if* $b = a$, *if* $a = b$ *and* $b = c$ *then* $a = c$.

2. *The sum of two objects a and b is defined, i.e. an object is defined denoted by* $a + b$, *also belonging to the system, which satisfies*

If $a = b$ *then* $a + c = b + c$, $a + b = b + a$, $a + (b + c) = (a + b) + c$, *and the common value of the last equality is denoted by* $a + b + c$.

what?   3. *If a is an object of the system and m a positive integer, then we understand by* ma *the sum of m objects equal to a. It is easy to see that for objects* $a, b, \ldots$ *of the system and positive integers* $m, n, \ldots$ *one has*

If $a = b$ *then* $ma = mb$, $m(a + b) = ma + mb$, $(m + n)a = ma + na$, $m(na) = mna$, $1a = a$.

We suppose that for any real number m the notation ma has a meaning such that the preceding equations are valid.

Peano also postulated the existence of a zero object 0 and used the notation $a - b$ for $a + (-b)$. By introducing the notions of dependent and independent objects, he defined the notion of dimension, showed that finite-dimensional spaces have a basis and gave examples of infinite-dimensional linear spaces.

If one considers only functions of degree n, then these functions form a linear system with $n + 1$ *dimensions, the entire functions of arbitrary degree form a linear system with infinitely many dimensions.*

Peano also introduced linear operators on a linear space and showed that by using coordinates one obtains a matrix.

With the passage of time, much concrete has set on these foundations. Techniques and notation have become more refined and the range of applications greatly enlarged. Nowadays *Linear Algebra*, comprising matrices and vector spaces, plays a major rôle in the mathematical curriculum. Notwithstanding the fact that many important and powerful computer packages exist to solve problems in linear algebra, it is our contention that a sound knowledge of the basic concepts and techniques is essential.

Contents

1.
The Algebra of Matrices

If m and n are positive integers then by a **matrix of size m by n**, or an $m \times n$ **matrix**, we shall mean a rectangular array consisting of mn numbers in a boxed display consisting of m rows and n columns. Simple examples of such objects are the following:

$$1 \times 5 : \begin{bmatrix} 10 & 9 & 8 & 7 & 6 \end{bmatrix} \qquad 3 \times 2 : \begin{bmatrix} 1 & 2 \\ 3 & 4 \\ 5 & 6 \end{bmatrix}$$

$$4 \times 4 : \begin{bmatrix} 1 & 2 & 3 & 4 \\ 2 & 3 & 4 & 5 \\ 3 & 4 & 5 & 6 \\ 4 & 5 & 6 & 7 \end{bmatrix} \qquad 3 \times 1 : \begin{bmatrix} 0 \\ 2 \\ 4 \end{bmatrix}$$

In general we shall display an $m \times n$ matrix as

$$\begin{bmatrix} x_{11} & x_{12} & x_{13} & \cdots & x_{1n} \\ x_{21} & x_{22} & x_{23} & \cdots & x_{2n} \\ x_{31} & x_{32} & x_{33} & \cdots & x_{3n} \\ \vdots & \vdots & \vdots & \ddots & \vdots \\ x_{m1} & x_{m2} & x_{m3} & \cdots & x_{mn} \end{bmatrix}.$$

- Note that the *first* suffix gives the number of the **row** and the *second* suffix that of the **column**, so that x_{ij} appears at the intersection of the i-th row and the j-th column.

We shall often find it convenient to abbreviate the above display to simply

$$[x_{ij}]_{m \times n}$$

and refer to x_{ij} as the (i,j)-**th element** or the (i,j)-**th entry** of the matrix.

- Thus the expression $X = [x_{ij}]_{m \times n}$ will be taken to mean that 'X is the $m \times n$ matrix whose (i,j)-th element is x_{ij}'.

Example 1.1

The 3×3 matrix $X = \begin{bmatrix} 1 & 1 & 1 \\ 2 & 2^2 & 2^3 \\ 3 & 3^2 & 3^3 \end{bmatrix}$ can be expressed as $X = [x_{ij}]_{3 \times 3}$ where $x_{ij} = i^j$.

Example 1.2

The 3×3 matrix $X = \begin{bmatrix} a & a & a \\ 0 & a & a \\ 0 & 0 & a \end{bmatrix}$ can be expressed as $X = [x_{ij}]_{3 \times 3}$ where

$$x_{ij} = \begin{cases} a & \text{if } i \leqslant j; \\ 0 & \text{otherwise.} \end{cases}$$

Example 1.3

The $n \times n$ matrix

$$X = \begin{bmatrix} 1 & 0 & 0 & \dots & 0 \\ e & 1 & 0 & \dots & 0 \\ e^2 & e & 1 & \dots & 0 \\ \vdots & \vdots & \vdots & \ddots & \vdots \\ e^{n-1} & e^{n-2} & e^{n-3} & \dots & 1 \end{bmatrix}$$

can be expressed as $X = [x_{ij}]_{n \times n}$ where

$$x_{ij} = \begin{cases} e^{i-j} & \text{if } i \geqslant j; \\ 0 & \text{otherwise.} \end{cases}$$

EXERCISES

1.1 Write out the 3×3 matrix whose entries are given by $x_{ij} = i + j$.

1.2 Write out the 3×3 matrix whose entries are given by

$$x_{ij} = \begin{cases} 1 & \text{if } i + j \text{ is even;} \\ 0 & \text{otherwise.} \end{cases}$$

1.3 Write out the 3×3 matrix whose entries are given by $x_{ij} = (-1)^{i-j}$.

1.4 Write out the $n \times n$ matrix whose entries are given by

$$x_{ij} = \begin{cases} -1 & \text{if } i > j; \\ 0 & \text{if } i = j; \\ 1 & \text{if } i < j. \end{cases}$$

1.5 Write out the 6×6 matrix $A = [a_{ij}]$ in which a_{ij} is given by

(1) the least common multiple of i and j;

(2) the greatest common divisor of i and j.

1.6 Given the $n \times n$ matrix $A = [a_{ij}]$, describe the $n \times n$ matrix $B = [b_{ij}]$ which is such that $b_{ij} = a_{i,n+1-j}$.

Before we can develop an algebra for matrices, it is essential that we decide what is meant by saying that two matrices are *equal*. Common sense dictates that this should happen only if the matrices in question are of the same size and have corresponding entries equal.

Definition

If $A = [a_{ij}]_{m \times n}$ and $B = [b_{ij}]_{p \times q}$ then we shall say that A and B are **equal** (and write $A = B$) if and only if

(1) $m = p$ and $n = q$;

(2) $a_{ij} = b_{ij}$ for all i, j.

The algebraic system that we shall develop for matrices will have many of the familiar properties enjoyed by the system of real numbers. However, as we shall see, there are some very striking differences.

Definition

Given $m \times n$ matrices $A = [a_{ij}]$ and $B = [b_{ij}]$, we define the **sum** $A + B$ to be the $m \times n$ matrix whose (i,j)-th element is $a_{ij} + b_{ij}$.

Note that the sum $A + B$ is defined only when A and B are of the same size; and to obtain this sum we simply add corresponding entries, thereby obtaining a matrix again of the same size. Thus, for instance,

$$\begin{bmatrix} -1 & 0 \\ 2 & -2 \end{bmatrix} + \begin{bmatrix} 1 & 2 \\ -1 & 0 \end{bmatrix} = \begin{bmatrix} 0 & 2 \\ 1 & -2 \end{bmatrix}.$$

Theorem 1.1

Addition of matrices is

(1) *commutative* [in the sense that if A, B are of the same size then we have $A + B = B + A$];

(2) *associative* [in the sense that if A, B, C are of the same size then we have $A + (B + C) = (A + B) + C$].

Proof

(1) If A and B are each of size $m \times n$ then $A + B$ and $B + A$ are also of size $m \times n$ and by the above definition we have

$$A + B = [a_{ij} + b_{ij}], \qquad B + A = [b_{ij} + a_{ij}].$$

Since addition of numbers is commutative we have $a_{ij} + b_{ij} = b_{ij} + a_{ij}$ for all i,j and so, by the definition of equality for matrices, we conclude that $A + B = B + A$.

(2) If A, B, C are each of size $m \times n$ then so are $A + (B + C)$ and $(A + B) + C$. Now the (i,j)-th element of $A + (B+C)$ is $a_{ij} + (b_{ij} + c_{ij})$ whereas that of $(A+B)+C$ is $(a_{ij} + b_{ij}) + c_{ij}$. Since addition of numbers is associative we have $a_{ij} + (b_{ij} + c_{ij}) = (a_{ij}+b_{ij})+c_{ij}$ for all i,j and so, by the definition of equality for matrices, we conclude that $A + (B + C) = (A + B) + C$. $\square$

Because of Theorem 1.1(2) we agree, as with numbers, to write $A + B + C$ for either $A + (B + C)$ or $(A + B) + C$.

Theorem 1.2

There is a unique $m \times n$ matrix M such that, for every $m \times n$ matrix A, $A + M = A$.

Proof

Consider the matrix $M = [m_{ij}]_{m \times n}$ all of whose entries are 0; i.e. $m_{ij} = 0$ for all i,j. For every matrix $A = [a_{ij}]_{m \times n}$ we have

$$A + M = [a_{ij} + m_{ij}]_{m \times n} = [a_{ij} + 0]_{m \times n} = [a_{ij}]_{m \times n} = A.$$

To establish the uniqueness of this matrix M, suppose that $B = [b_{ij}]_{m \times n}$ is also such that $A + B = A$ for every $m \times n$ matrix A. Then in particular we have $M + B = M$. But, taking B instead of A in the property for M, we have $B + M = B$. It now follows by Theorem 1.1(1) that $B = M$. $\square$

Definition

The unique matrix arising in Theorem 1.2 is called the $m \times n$ **zero matrix** and will be denoted by $0_{m \times n}$, or simply by 0 if no confusion arises.

Theorem 1.3

For every $m \times n$ matrix A there is a unique $m \times n$ matrix B such that $A + B = 0$.

Proof

Given $A = [a_{ij}]_{m \times n}$, consider $B = [-a_{ij}]_{m \times n}$, i.e. the matrix whose (i,j)-th element is the additive inverse of the (i,j)-th element of A. Clearly, we have

$$A + B = [a_{ij} + (-a_{ij})]_{m \times n} = 0.$$

To establish the uniqueness of such a matrix B, suppose that $C = [c_{ij}]_{m \times n}$ is also such that $A + C = 0$. Then for all i,j we have $a_{ij} + c_{ij} = 0$ and consequently $c_{ij} = -a_{ij}$ which means, by the above definition of equality, that $C = B$. $\square$

Definition

The unique matrix B arising in Theorem 1.3 is called the **additive inverse** of A and will be denoted by $-A$. Thus $-A$ is the matrix whose elements are the additive inverses of the corresponding elements of A.

Given numbers x, y the **difference** $x - y$ is defined to be $x + (-y)$. For matrices A, B of the same size we shall similarly write $A - B$ for $A + (-B)$, the operation '$-$' so defined being called **subtraction** of matrices.

EXERCISES

1.7 Show that subtraction of matrices is neither commutative nor associative.

1.8 Prove that if A and B are of the same size then $-(A + B) = -A - B$.

1.9 Simplify $\begin{bmatrix} x - y & y - z & z - w \\ w - x & x - y & y - z \end{bmatrix} - \begin{bmatrix} x - w & y - x & z - y \\ y - x & z - y & w - z \end{bmatrix}$.

So far our matrix algebra has been confined to the operation of addition. This is a simple extension of the same notion for numbers, for we can think of 1×1 matrices as behaving essentially as numbers. We shall now investigate how the notion of multiplication for numbers can be extended to matrices. This, however, is not quite so straightforward. There are in fact two distinct 'multiplications' that can be defined. The first 'multiplies' a matrix by a number, and the second 'multiplies' a matrix by another matrix.

Definition

Given a matrix A and a number λ we define the **product of A by λ** to be the matrix, denoted by λA, that is obtained from A by multiplying every element of A by λ. Thus, if $A = [a_{ij}]_{m \times n}$ then $\lambda A = [\lambda a_{ij}]_{m \times n}$.

This operation is traditionally called **multiplying a matrix by a scalar** (where the word *scalar* is taken to be synonymous with *number*). Such multiplication by scalars may also be thought of as scalars *acting* on matrices. The principal properties of this action are as follows.

Theorem 1.4

If A and B are $m \times n$ matrices then, for any scalars λ and μ,

(1) $\lambda(A + B) = \lambda A + \lambda B$;
(2) $(\lambda + \mu)A = \lambda A + \mu A$;
(3) $\lambda(\mu A) = (\lambda \mu)A$;
(4) $(-1)A = -A$;
(5) $0A = 0_{m \times n}$.

Proof

Let $A = [a_{ij}]_{m \times n}$ and $B = [b_{ij}]_{m \times n}$. Then the above equalities follow from the observations

(1) $\lambda(a_{ij} + b_{ij}) = \lambda a_{ij} + \lambda b_{ij}$.
(2) $(\lambda + \mu)a_{ij} = \lambda a_{ij} + \mu a_{ij}$.

(3) $\lambda(\mu a_{ij}) = (\lambda\mu)a_{ij}$.

(4) $(-1)a_{ij} = -a_{ij}$.

(5) $0a_{ij} = 0$. □

Observe that for every positive integer n we have

$$nA = A + A + \cdots + A \qquad (n \text{ terms}).$$

This follows immediately from the definition of the product λA; for the (i,j)-th element of nA is $na_{ij} = a_{ij} + a_{ij} + \cdots + a_{ij}$, there being n terms in the summation.

EXERCISES

1.10 Given any $m \times n$ matrices A and B, solve the matrix equation

$$3(X + \tfrac{1}{2}A) = 5(X - \tfrac{3}{4}B).$$

1.11 Given the matrices

$$A = \begin{bmatrix} 1 & 0 & 0 \\ 0 & 1 & 0 \\ 0 & 0 & 1 \end{bmatrix}, \qquad B = \begin{bmatrix} 1 & 1 & 1 \\ 1 & 1 & 1 \\ 1 & 1 & 1 \end{bmatrix}$$

solve the matrix equation $X + A = 2(X - B)$.

We shall now describe the operation that is called **matrix multiplication**. This is the 'multiplication' of one matrix by another. At first sight this concept (due originally to Cayley) appears to be a most curious one. Whilst it has in fact a very natural interpretation in an algebraic context that we shall see later, we shall for the present simply accept it without asking how it arises. Having said this, however, we shall illustrate its importance in Chapter 2, particularly in the applications of matrix algebra.

Definition

Let $A = [a_{ij}]_{m \times n}$ and $B = [b_{ij}]_{n \times p}$ (note the sizes!). Then we define the **product** AB to be the $m \times p$ matrix whose (i,j)-th element is

$$[AB]_{ij} = a_{i1}b_{1j} + a_{i2}b_{2j} + a_{i3}b_{3j} + \cdots + a_{in}b_{nj}.$$

In other words, the (i,j)-th element of the product AB is obtained by summing the products of the elements in the i-th **row** of A with the corresponding elements in the j-th **column** of B.

The above expression for $[AB]_{ij}$ can be written in abbreviated form using the so-called Σ **-notation**:

$$[AB]_{ij} = \sum_{k=1}^{n} a_{ik}b_{kj}.$$

The process for computing products can be pictorially summarised as follows :

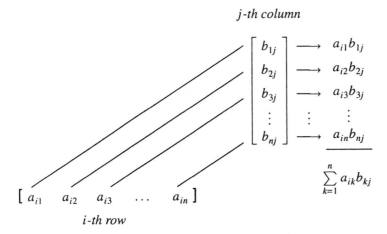

j-th column

It is important to note that, in forming these sums of products, there are no elements that are 'left over' since in the definition of the product AB the number n of columns of A is the same as the number of rows of B.

Example 1.4

Consider the matrices

$$A = \begin{bmatrix} 0 & 1 & 0 \\ 2 & 3 & 1 \end{bmatrix}, \qquad B = \begin{bmatrix} 2 & 0 \\ 1 & 2 \\ 1 & 1 \end{bmatrix}.$$

The product AB is defined since A is of size $2 \times \mathbf{3}$ and B is of size $\mathbf{3} \times 2$; moreover, AB is of size 2×2. We have

$$AB = \begin{bmatrix} 0{\cdot}2 + 1{\cdot}1 + 0{\cdot}1 & 0{\cdot}0 + 1{\cdot}2 + 0{\cdot}1 \\ 2{\cdot}2 + 3{\cdot}1 + 1{\cdot}1 & 2{\cdot}0 + 3{\cdot}2 + 1{\cdot}1 \end{bmatrix} = \begin{bmatrix} 1 & 2 \\ 8 & 7 \end{bmatrix}.$$

Note that in this case the product BA is also defined (since B has the same number of columns as A has rows). The product BA is of size 3×3 :

$$BA = \begin{bmatrix} 2{\cdot}0 + 0{\cdot}2 & 2{\cdot}1 + 0{\cdot}3 & 2{\cdot}0 + 0{\cdot}1 \\ 1{\cdot}0 + 2{\cdot}2 & 1{\cdot}1 + 2{\cdot}3 & 1{\cdot}0 + 2{\cdot}1 \\ 1{\cdot}0 + 1{\cdot}2 & 1{\cdot}1 + 1{\cdot}3 & 1{\cdot}0 + 1{\cdot}1 \end{bmatrix} = \begin{bmatrix} 0 & 2 & 0 \\ 4 & 7 & 2 \\ 2 & 4 & 1 \end{bmatrix}.$$

The above example exhibits a curious fact concerning matrix multiplication, namely that if AB and BA are defined then these products need not be equal. Indeed, as we have just seen, AB and BA need not even be of the same size. It is also possible for AB and BA to be defined and of the same size and still be not equal:

Example 1.5

The matrices

$$A = \begin{bmatrix} 0 & 1 \\ 0 & 0 \end{bmatrix}, \qquad B = \begin{bmatrix} 1 & 0 \\ 0 & 0 \end{bmatrix}$$

are such that $AB = 0$ and $BA = A$.

We thus observe that *in general matrix multiplication is not commutative*.

EXERCISES

1.12 Compute the matrix product

$$\begin{bmatrix} 3 & 1 & -2 \\ 2 & -2 & 0 \\ -1 & 1 & 2 \end{bmatrix} \begin{bmatrix} 1 & 1 & 1 \\ 1 & -1 & 1 \\ 0 & 1 & 2 \end{bmatrix}.$$

1.13 Compute the matrix product

$$\begin{bmatrix} 1 & 1 & 1 \\ 0 & 1 & 1 \\ 0 & 0 & 1 \end{bmatrix} \begin{bmatrix} 1 & 0 & 0 \\ 1 & 1 & 0 \\ 1 & 1 & 1 \end{bmatrix}.$$

1.14 Compute the matrix products

$$\begin{bmatrix} 1 \\ 2 \\ 3 \\ 4 \end{bmatrix} \begin{bmatrix} 1 & 2 & 3 & 4 \end{bmatrix}, \qquad \begin{bmatrix} 1 & 2 & 3 & 4 \end{bmatrix} \begin{bmatrix} 1 \\ 2 \\ 3 \\ 4 \end{bmatrix}.$$

1.15 Given the matrices

$$A = \begin{bmatrix} 3 & 0 \\ -1 & 2 \\ 1 & 1 \end{bmatrix}, \qquad B = \begin{bmatrix} 4 & -1 \\ 0 & 2 \end{bmatrix}, \qquad C = \begin{bmatrix} 1 & 4 & 2 \\ 3 & 1 & 5 \end{bmatrix},$$

compute the products $(AB)C$ and $A(BC)$.

We now consider the basic properties of matrix multiplication.

Theorem 1.5

Matrix multiplication is associative [in the sense that, when the products are defined, $A(BC) = (AB)C$].

Proof

For $A(BC)$ to be defined we require the respective sizes to be $m \times n$, $n \times p$, $p \times q$ in which case the product $A(BC)$ is also defined, and conversely. Computing the

(i,j)-th element of this product, we obtain

$$[A(BC)]_{ij} = \sum_{k=1}^{n} a_{ik}[BC]_{kj} = \sum_{k=1}^{n} a_{ik}\left(\sum_{t=1}^{p} b_{kt}c_{tj}\right)$$

$$= \sum_{k=1}^{n} \sum_{t=1}^{p} a_{ik}b_{kt}c_{tj}.$$

If we now compute the (i,j)-th element of $(AB)C$, we obtain the same:

$$[(AB)C]_{ij} = \sum_{t=1}^{p}[AB]_{it}c_{tj} = \sum_{t=1}^{p}\left(\sum_{k=1}^{n} a_{ik}b_{kt}\right)c_{tj}$$

$$= \sum_{t=1}^{p} \sum_{k=1}^{n} a_{ik}b_{kt}c_{tj}.$$

Consequently we see that $A(BC) = (AB)C$. $\square$

Because of Theorem 1.5 we shall write ABC for either $A(BC)$ or $(AB)C$. Also, for every positive integer n we shall write A^n for the product $AA \cdots A$ (n terms).

EXERCISES

1.16 Compute the matrix product

$$\begin{bmatrix} x & y & 1 \end{bmatrix} \begin{bmatrix} a & h & g \\ h & b & f \\ g & f & c \end{bmatrix} \begin{bmatrix} x \\ y \\ 1 \end{bmatrix}.$$

Hence express in matrix notation the equations

(1) $x^2 + 9xy + y^2 + 8x + 5y + 2 = 0$;

(2) $\dfrac{x^2}{\alpha^2} + \dfrac{y^2}{\beta^2} = 1$;

(3) $xy = \alpha^2$;

(4) $y^2 = 4\alpha x$.

1.17 Compute A^2 and A^3 where $A = \begin{bmatrix} 0 & a & a^2 \\ 0 & 0 & a \\ 0 & 0 & 0 \end{bmatrix}$.

Matrix multiplication and matrix addition are connected by the following **distributive laws**.

Theorem 1.6

When the relevant sums and products are defined, we have

$$A(B + C) = AB + AC, \qquad (B + C)A = BA + CA.$$

Proof

For the first equality we require A to be of size $m \times n$ and B, C to be of size $n \times p$, in which case

$$[A(B+C)]_{ij} = \sum_{k=1}^{n} a_{ik}[B+C]_{kj} = \sum_{k=1}^{n} a_{ik}(b_{kj} + c_{kj})$$
$$= \sum_{k=1}^{n} a_{ik}b_{kj} + \sum_{k=1}^{n} a_{ik}c_{kj}$$
$$= [AB]_{ij} + [AC]_{ij}$$
$$= [AB + AC]_{ij}$$

and it follows that $A(B+C) = AB + AC$.

For the second equality, in which we require B, C to be of size $m \times n$ and A to be of size $n \times p$, a similar argument applies. □

Matrix multiplication is also connected with multiplication by scalars.

Theorem 1.7

If AB is defined then for all scalars λ we have

$$\lambda(AB) = (\lambda A)B = A(\lambda B).$$

Proof

The (i, j)-th elements of the three mixed products are

$$\lambda\Big(\sum_{k=1}^{n} a_{ik}b_{kj}\Big) = \sum_{k=1}^{n}(\lambda a_{ik})b_{kj} = \sum_{k=1}^{n} a_{ik}(\lambda b_{kj}),$$

from which the result follows. □

EXERCISES

1.18 Consider the matrices

$$A = \begin{bmatrix} 0 & 1 \\ 0 & 1 \end{bmatrix}, \qquad B = \begin{bmatrix} -1 & -1 \\ 0 & 0 \end{bmatrix}.$$

Prove that

$$(A+B)^2 \neq A^2 + 2AB + B^2,$$

but that

$$(A+B)^3 = A^3 + 3A^2B + 3AB^2 + B^3.$$

Definition

A matrix is said to be **square** if it is of size $n \times n$; i.e. has the same number of rows and columns.

Our next result is the multiplicative analogue of Theorem 1.2, but the reader should note that it applies only in the case of square matrices.

Theorem 1.8

There is a unique $n \times n$ matrix M with the property that, for every $n \times n$ matrix A, $AM = A = MA$.

Proof

Consider the $n \times n$ matrix

$$M = \begin{bmatrix} 1 & 0 & 0 & \dots & 0 \\ 0 & 1 & 0 & \dots & 0 \\ 0 & 0 & 1 & \dots & 0 \\ \vdots & \vdots & \vdots & \ddots & \vdots \\ 0 & 0 & 0 & \dots & 1 \end{bmatrix}.$$

More precisely, if we define the **Kronecker symbol** δ_{ij} by

$$\delta_{ij} = \begin{cases} 1 & \text{if } i = j; \\ 0 & \text{otherwise,} \end{cases}$$

then we have $M = [\delta_{ij}]_{n \times n}$. If $A = [a_{ij}]_{n \times n}$ then $[AM]_{ij} = \sum_{k=1}^{n} a_{ik}\delta_{kj} = a_{ij}$, the last equality following from the fact that every term in the summation is 0 except that in which $k = j$, and this term is $a_{ij}1 = a_{ij}$. We deduce, therefore, that $AM = A$. Similarly, we can show that $MA = A$. This then establishes the existence of a matrix M with the stated property.

To show that such a matrix M is unique, suppose that P is also an $n \times n$ matrix such that $AP = A = PA$ for every $n \times n$ matrix A. Then in particular we have $MP = M = PM$. But, by the same property for M, we have $PM = P = MP$. Thus we see that $P = M$. $\square$

Definition

The unique matrix M described in Theorem 1.8 is called the $n \times n$ **identity matrix** and will be denoted by I_n.

Note that I_n has all of its 'diagonal' entries equal to 1 and all other entries 0. This is a special case of the following important type of square matrix.

Definition

A square matrix $D = [d_{ij}]_{n \times n}$ is said to be **diagonal** if $d_{ij} = 0$ whenever $i \neq j$. Less formally, D is diagonal when all the entries off the main diagonal are 0.

EXERCISES

1.19 If A and B are $n \times n$ diagonal matrices prove that so also is AB.

1.20 If A and B are $n \times n$ diagonal matrices prove that so also is $A^p B^q$ for all positive integers p, q.

There is no multiplicative analogue of Theorem 1.3; for example, if

$$A = \begin{bmatrix} 0 & 1 \\ 0 & 0 \end{bmatrix}$$

then we have

$$\begin{bmatrix} a & b \\ c & d \end{bmatrix} \begin{bmatrix} 0 & 1 \\ 0 & 0 \end{bmatrix} = \begin{bmatrix} 0 & a \\ 0 & c \end{bmatrix},$$

so there is no matrix M such that $MA = I_2$.

Note also that several of the familiar laws of high-school algebra break down in this new algebraic system. This is illustrated in particular in Example 1.5 and Exercise 1.14 above.

Definition

We say that matrices A, B **commute** if $AB = BA$. Note that for A, B to commute it is necessary that they be square and of the same size.

EXERCISES

1.21 If A and B commute, prove that so also do A^m and B^n for all positive integers m and n.

1.22 Using an inductive argument, prove that if A and B commute then the usual binomial expansion is valid for $(A + B)^n$.

There are other curious properties of matrix multiplication. We mention in particular the following examples, which illustrate in a very simple way the fact that matrix multiplication has to be treated with some care.

Example 1.6

If λ is a non-zero real number then the equality

$$\begin{bmatrix} 0 & \lambda \\ \lambda^{-1} & 0 \end{bmatrix} \begin{bmatrix} 0 & \lambda \\ \lambda^{-1} & 0 \end{bmatrix} = I_2$$

shows that I_2 has infinitely many square roots!

Example 1.7

It follows by Theorem 1.4(5) that the matrix $0_{n \times n}$ has the property that $0_{n \times n} A = 0_{n \times n} = A 0_{n \times n}$ for every $n \times n$ matrix A. In Example 1.5 we have seen that it is possible for the product of two non-zero matrices to be the zero matrix.

Definition

If A is an $m \times n$ matrix then by the **transpose** of A we mean the $n \times m$ matrix A' whose (i, j)-th element is the (j, i)-th element of A. More precisely, if $A = [a_{ij}]_{m \times n}$ then $A' = [a_{ji}]_{n \times m}$.

The principal properties of transposition of matrices are summarised in the following result.

Theorem 1.9

When the relevant sums and products are defined, we have

$$(A')' = A, \quad (A + B)' = A' + B', \quad (\lambda A)' = \lambda A', \quad (AB)' = B'A'.$$

Proof

The first three equalities follow immediately from the definitions. To prove that $(AB)' = B'A'$ (note the reversal!), suppose that $A = [a_{ij}]_{m \times n}$ and $B = [b_{ij}]_{n \times p}$. Then $(AB)'$ and $B'A'$ are each of size $p \times m$. Since

$$[B'A']_{ij} = \sum_{k=1}^{n} b_{ki} a_{jk} = \sum_{k=1}^{n} a_{jk} b_{ki} = [AB]_{ji}$$

we deduce that $(AB)' = B'A'$. $\square$

EXERCISES

1.23 Prove that, when either expression is meaningful,

$$[A(B + C)]' = B'A' + C'A'.$$

1.24 Prove by induction that $(A^n)' = (A')^n$ for every positive integer n.

1.25 If A and B commute, prove that so also do A' and B'.

1.26 Let $\mathbf{X} = [a \ b \ c]$ and

$$A = \begin{bmatrix} 0 & -c & b \\ c & 0 & -a \\ -b & a & 0 \end{bmatrix}$$

where $a^2 + b^2 + c^2 = 1$.

(1) Show that $A^2 = \mathbf{X}'\mathbf{X} - I_2$.

(2) Prove that $A^3 = -A$.

(3) Find A^4 in terms of $\mathbf{X}$.

Definition

A square matrix A is **symmetric** if $A = A'$; and **skew-symmetric** if $A = -A'$.

Example 1.8

For every square matrix A the matrix $A + A'$ is symmetric, and the matrix $A - A'$ is skew-symmetric. In fact, by Theorem 1.9, we have

$$(A + A')' = A' + (A')' = A' + A;$$
$$(A - A')' = A' - (A')' = A' - A = -(A - A').$$

Theorem 1.10

Every square matrix can be expressed uniquely as the sum of a symmetric matrix and a skew-symmetric matrix.

Proof

The equality

$$A = \tfrac{1}{2}(A + A') + \tfrac{1}{2}(A - A'),$$

together with Example 1.8, shows that such an expression exists. As for uniqueness, suppose that $A = B + C$ where B is symmetric and C is skew-symmetric. Then $A' = B' + C' = B - C$. It follows from these equations that $B = \tfrac{1}{2}(A + A')$ and $C = \tfrac{1}{2}(A - A')$. $\square$

EXERCISES

1.27 Prove that the zero square matrices are the only matrices that are both symmetric and skew-symmetric.

1.28 Let A, B be of size $n \times n$ with A symmetric and B skew-symmetric. Determine which of the following are symmetric and which are skew-symmetric:

$$AB + BA, \quad AB - BA, \quad A^2, \quad B^2, \quad A^p B^q A^p \ (p, q \text{ positive integers}).$$

SUPPLEMENTARY EXERCISES

1.29 Let $\mathbf{x}$ and $\mathbf{y}$ be $n \times 1$ matrices. Show that the matrix

$$A = \mathbf{xy'} - \mathbf{yx'}$$

is of size $n \times n$ and is skew-symmetric. Show also that $\mathbf{x'y}$ and $\mathbf{y'x}$ are of size 1×1 and are equal.

If $\mathbf{x'x} = \mathbf{y'y} = [1]$ and $\mathbf{x'y} = \mathbf{y'x} = [k]$, prove that $A^3 = (k^2 - 1)A$.

1.30 Show that if A and B are 2×2 matrices then the sum of the diagonal elements of $AB - BA$ is zero.

If E is a 2×2 matrix and the sum of the diagonal elements of E is zero, show that $E^2 = \lambda I_2$ for some scalar λ.

Deduce from the above that if A, B, C are 2×2 matrices then

$$(AB - BA)^2 C = C(AB - BA)^2.$$

1.31 Determine all 2×2 matrices X with real entries such that $X^2 = I_2$.

1.32 Show that there are infinitely many real 2×2 matrices A with $A^2 = -I_2$.

1.33 Let A be the matrix

$$\begin{bmatrix} 0 & a & a^2 & a^3 \\ 0 & 0 & a & a^2 \\ 0 & 0 & 0 & a \\ 0 & 0 & 0 & 0 \end{bmatrix}.$$

Define the matrix B by

$$B = A - \tfrac{1}{2}A^2 + \tfrac{1}{3}A^3 - \tfrac{1}{4}A^4 + \cdots.$$

Show that this series has only finitely many terms different from zero and calculate B. Show also that the series

$$B + \tfrac{1}{2!}B^2 + \tfrac{1}{3!}B^3 + \cdots$$

has only finitely many non-zero terms and that its sum is A.

1.34 If $A = \begin{bmatrix} \cos \vartheta & \sin \vartheta \\ -\sin \vartheta & \cos \vartheta \end{bmatrix}$ and $B = \begin{bmatrix} \cos \varphi & \sin \varphi \\ -\sin \varphi & \cos \varphi \end{bmatrix}$ prove that

$$AB = \begin{bmatrix} \cos(\vartheta + \varphi) & \sin(\vartheta + \varphi) \\ -\sin(\vartheta + \varphi) & \cos(\vartheta + \varphi) \end{bmatrix}.$$

1.35 If $A = \begin{bmatrix} \cos \vartheta & \sin \vartheta \\ -\sin \vartheta & \cos \vartheta \end{bmatrix}$ prove that $A^n = \begin{bmatrix} \cos n\vartheta & \sin n\vartheta \\ -\sin n\vartheta & \cos n\vartheta \end{bmatrix}.$

1.36 Prove that, for every positive integer n,

$$\begin{bmatrix} \alpha & 1 & 0 \\ 0 & \alpha & 1 \\ 0 & 0 & \alpha \end{bmatrix}^n = \begin{bmatrix} \alpha^n & n\alpha^{n-1} & \tfrac{1}{2}n(n-1)\alpha^{n-2} \\ 0 & \alpha^n & n\alpha^{n-1} \\ 0 & 0 & \alpha^n \end{bmatrix}.$$

1.37 If A and B are $n \times n$ matrices, define the **Lie product**

$$[AB] = AB - BA.$$

Establish the following identities :

(1) $[[AB]C] + [[BC]A] + [[CA]B] = 0$;

(2) $[(A + B)C] = [AC] + [BC]$;

(3) $[[[AB]C]D] + [[[BC]D]A] + [[[CD]A]B] + [[[DA]B]C] = 0$.

Show by means of an example that in general $[[AB]C] \neq [A[BC]]$.

1.38 Given that $x = \dfrac{a_1 y + a_2}{a_3 y + a_4}$ and $y = \dfrac{b_1 z + b_2}{b_3 z + b_4}$ prove that $x = \dfrac{c_1 z + c_2}{c_3 z + c_4}$

where

$$\begin{bmatrix} c_1 & c_2 \\ c_3 & c_4 \end{bmatrix} = \begin{bmatrix} a_1 & a_2 \\ a_3 & a_4 \end{bmatrix} \begin{bmatrix} b_1 & b_2 \\ b_3 & b_4 \end{bmatrix}.$$

2.

Some Applications of Matrices

We shall now give brief descriptions of some situations to which matrix theory finds a natural application, and some problems to which the solutions are determined by the algebra that we have developed. Some of these applications will be dealt with in greater detail in later chapters.

1. Analytic geometry

In analytic geometry, various transformations of the coordinate axes may be described using matrices. By way of example, suppose that in the two-dimensional cartesian plane we rotate the coordinate axes in an anti-clockwise direction through an angle ϑ, as illustrated in the following diagram:

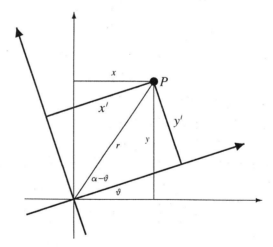

Let us compute the new coordinates (x', y') of the point P whose old coordinates were (x, y).

From the diagram we have $x = r \cos \alpha$ and $y = r \sin \alpha$ so

$$x' = r \cos(\alpha - \vartheta) = r \cos \alpha \cos \vartheta + r \sin \alpha \sin \vartheta$$
$$= x \cos \vartheta + y \sin \vartheta;$$

$$y' = r \sin(\alpha - \vartheta) = r \sin \alpha \cos \vartheta - r \cos \alpha \sin \vartheta$$
$$= y \cos \vartheta - x \sin \vartheta.$$

These equations give x', y' in terms of x, y and ϑ. They can be expressed in the matrix form

$$\begin{bmatrix} x' \\ y' \end{bmatrix} = \begin{bmatrix} \cos \vartheta & \sin \vartheta \\ -\sin \vartheta & \cos \vartheta \end{bmatrix} \begin{bmatrix} x \\ y \end{bmatrix}.$$

The 2×2 matrix

$$R_\vartheta = \begin{bmatrix} \cos \vartheta & \sin \vartheta \\ -\sin \vartheta & \cos \vartheta \end{bmatrix}$$

is called the **rotation matrix** associated with ϑ. It has the following property:

$$R_\vartheta R'_\vartheta = I_2 = R'_\vartheta R_\vartheta.$$

In fact, we have

$$R_\vartheta R'_\vartheta = \begin{bmatrix} \cos \vartheta & \sin \vartheta \\ -\sin \vartheta & \cos \vartheta \end{bmatrix} \begin{bmatrix} \cos \vartheta & -\sin \vartheta \\ \sin \vartheta & \cos \vartheta \end{bmatrix}$$
$$= \begin{bmatrix} \cos^2 \vartheta + \sin^2 \vartheta & -\cos \vartheta \sin \vartheta + \sin \vartheta \cos \vartheta \\ -\sin \vartheta \cos \vartheta + \cos \vartheta \sin \vartheta & \sin^2 \vartheta + \cos^2 \vartheta \end{bmatrix}$$
$$= \begin{bmatrix} 1 & 0 \\ 0 & 1 \end{bmatrix},$$

and similarly, as the reader can verify, $R'_\vartheta R_\vartheta = I_2$.

This leads us more generally to the following notion.

Definition

An $n \times n$ matrix A is said to be **orthogonal** if

$$AA' = I_n = A'A.$$

It follows from the above that to every rotation of axes in two dimensions we can associate a real orthogonal matrix ('real' in the sense that its elements are real numbers).

EXERCISES

2.1 If A is an orthogonal $n \times n$ matrix prove that A' is also orthogonal.

2.2 If A and B are orthogonal $n \times n$ matrices prove that AB is also orthogonal.

2.3 Prove that a real 2×2 matrix is orthogonal if and only if it is of one of the forms

$$\begin{bmatrix} a & b \\ -b & a \end{bmatrix}, \quad \begin{bmatrix} a & b \\ b & -a \end{bmatrix}$$

where $a^2 + b^2 = 1$.

Consider now the effect of one rotation followed by another. Suppose that we transform (x, y) into (x', y') by a rotation through ϑ, then (x', y') into (x'', y'') by a rotation through φ. Then we have

$$\begin{bmatrix} x'' \\ y'' \end{bmatrix} = \begin{bmatrix} \cos \varphi & \sin \varphi \\ -\sin \varphi & \cos \varphi \end{bmatrix} \begin{bmatrix} x' \\ y' \end{bmatrix}$$

$$= \begin{bmatrix} \cos \varphi & \sin \varphi \\ -\sin \varphi & \cos \varphi \end{bmatrix} \begin{bmatrix} \cos \vartheta & \sin \vartheta \\ -\sin \vartheta & \cos \vartheta \end{bmatrix} \begin{bmatrix} x \\ y \end{bmatrix}.$$

This suggests that the effect of one rotation followed by another can be described by the product of the corresponding rotation matrices. Now it is intuitively clear that the order in which we perform the rotations does not matter, the final frame of reference being the same whether we first rotate through ϑ and then through φ or whether we rotate first through φ and then through ϑ. Intuitively, therefore, we can assert that *rotation matrices commute*. That this is indeed the case follows from the identities

$$R_\vartheta R_\varphi = R_{\vartheta+\varphi} = R_\varphi R_\vartheta$$

which the reader can easily verify as an exercise using the standard identities for $\cos(\vartheta + \varphi)$ and $\sin(\vartheta + \varphi)$.

Example 2.1

Consider the hyperbola whose equation is $x^2 - y^2 = 1$. If this is rotated through $45°$ anti-clockwise about the origin, what does its new equation become?

To answer this, observe first that rotating the hyperbola anti-clockwise through $45°$ is equivalent to rotating the axes clockwise through $45°$. Thus we have

$$\begin{bmatrix} x' \\ y' \end{bmatrix} = R_{-\pi/4} \begin{bmatrix} x \\ y \end{bmatrix}.$$

Now since

$$R_{\pi/4} R_{-\pi/4} = R_{\pi/4-\pi/4} = R_0 = I_2$$

we can multiply the above equation by $R_{\pi/4}$ to obtain

$$\begin{bmatrix} x \\ y \end{bmatrix} = R_{\pi/4} \begin{bmatrix} x' \\ y' \end{bmatrix} = \begin{bmatrix} \frac{1}{\sqrt{2}} & \frac{1}{\sqrt{2}} \\ -\frac{1}{\sqrt{2}} & \frac{1}{\sqrt{2}} \end{bmatrix} \begin{bmatrix} x' \\ y' \end{bmatrix}$$

so that

$$x = \tfrac{1}{\sqrt{2}} x' + \tfrac{1}{\sqrt{2}} y', \quad y = -\tfrac{1}{\sqrt{2}} x' + \tfrac{1}{\sqrt{2}} y'.$$

Thus the equation $x^2 - y^2 = 1$ transforms to

$$\left(\tfrac{1}{\sqrt{2}}x' + \tfrac{1}{\sqrt{2}}y'\right)^2 - \left(-\tfrac{1}{\sqrt{2}}x' + \tfrac{1}{\sqrt{2}}y'\right)^2 = 1,$$

i.e. to $2x'y' = 1$.

EXERCISES

2.4 Two similar sheets of graph paper are pinned together at the origin and the sheets are rotated. If the point $(1,0)$ of the top sheet is directly above the point $(\tfrac{5}{13}, \tfrac{12}{13})$ of the bottom sheet, above what point of the bottom sheet does the point $(2,3)$ of the top sheet lie?

2.5 For every point (x, y) of the cartesian plane let (x', y') be its reflection in the x-axis. Prove that

$$\begin{bmatrix} x' \\ y' \end{bmatrix} = \begin{bmatrix} 1 & 0 \\ 0 & -1 \end{bmatrix} \begin{bmatrix} x \\ y \end{bmatrix}.$$

2.6 In the cartesian plane let L be a line passing through the origin and making an angle ϑ with the x-axis. For every point (x, y) of the plane let (x_L, y_L) be its reflection in the line L. Prove that

$$\begin{bmatrix} x_L \\ y_L \end{bmatrix} = \begin{bmatrix} \cos 2\vartheta & \sin 2\vartheta \\ \sin 2\vartheta & -\cos 2\vartheta \end{bmatrix} \begin{bmatrix} x \\ y \end{bmatrix}.$$

2.7 In the cartesian plane let L be a line passing through the origin and making an angle ϑ with the x-axis. For every point (x, y) of the plane let $(x^\star, y^\star)$ be the projection of (x, y) onto L (i.e. the foot of the perpendicular from (x, y) to L). Prove that

$$\begin{bmatrix} x^\star \\ y^\star \end{bmatrix} = \begin{bmatrix} \cos^2 \vartheta & \sin \vartheta \cos \vartheta \\ \sin \vartheta \cos \vartheta & \sin^2 \vartheta \end{bmatrix} \begin{bmatrix} x \\ y \end{bmatrix}.$$

2. Systems of linear equations

As we have seen above, a pair of equations of the form

$$a_{11}x_1 + a_{12}x_2 = b_1, \quad a_{21}x_1 + a_{22}x_2 = b_2$$

can be expressed in the matrix form $A\mathbf{x} = \mathbf{b}$ where

$$A = \begin{bmatrix} a_{11} & a_{12} \\ a_{21} & a_{22} \end{bmatrix}, \quad \mathbf{x} = \begin{bmatrix} x_1 \\ x_2 \end{bmatrix}, \quad \mathbf{b} = \begin{bmatrix} b_1 \\ b_2 \end{bmatrix}.$$

Let us now consider the general case.

Definition

By a **system of m linear equations in the n unknowns** $x_1, \ldots, x_n$ we shall mean a list of equations of the form

$$a_{11}x_1 + a_{12}x_2 + a_{13}x_3 + \cdots + a_{1n}x_n = b_1$$
$$a_{21}x_1 + a_{22}x_2 + a_{23}x_3 + \cdots + a_{2n}x_n = b_2$$
$$a_{31}x_1 + a_{32}x_2 + a_{33}x_3 + \cdots + a_{3n}x_n = b_3$$
$$\vdots$$
$$a_{m1}x_1 + a_{m2}x_2 + a_{m3}x_3 + \cdots + a_{mn}x_n = b_m$$

where the a_{ij} and the b_i are numbers.

Since clearly

$$\begin{bmatrix} a_{11} & a_{12} & a_{13} & \cdots & a_{1n} \\ a_{21} & a_{22} & a_{23} & \cdots & a_{2n} \\ a_{31} & a_{32} & a_{33} & \cdots & a_{3n} \\ \vdots & \vdots & \vdots & \ddots & \vdots \\ a_{m1} & a_{m2} & a_{m3} & \cdots & a_{mn} \end{bmatrix} \begin{bmatrix} x_1 \\ x_2 \\ x_3 \\ \vdots \\ x_n \end{bmatrix} = \begin{bmatrix} a_{11}x_1 + a_{12}x_2 + a_{13}x_3 + \cdots + a_{1n}x_n \\ a_{21}x_1 + a_{22}x_2 + a_{23}x_3 + \cdots + a_{2n}x_n \\ a_{31}x_1 + a_{32}x_2 + a_{33}x_3 + \cdots + a_{3n}x_n \\ \vdots \\ a_{m1}x_1 + a_{m2}x_2 + a_{m3}x_3 + \cdots + a_{mn}x_n \end{bmatrix}$$

we see that this system can be expressed succinctly in the matrix form $A\mathbf{x} = \mathbf{b}$ where $A = [a_{ij}]_{m \times n}$ and $\mathbf{x}, \mathbf{b}$ are the column matrices

$$\mathbf{x} = \begin{bmatrix} x_1 \\ x_2 \\ \vdots \\ x_n \end{bmatrix}, \quad \mathbf{b} = \begin{bmatrix} b_1 \\ b_2 \\ \vdots \\ b_m \end{bmatrix}.$$

The $m \times n$ matrix A is called the **coefficient matrix** of the system. Note that it transforms a column matrix of length n into a column matrix of length m.

In the case where $\mathbf{b} = \mathbf{0}$ (i.e. where every $b_i = 0$) we say that the system is **homogeneous**.

If we adjoin to A the column $\mathbf{b}$ then we obtain an $m \times (n + 1)$ matrix which we write as $A|\mathbf{b}$ and call the **augmented matrix** of the system.

Whether or not a given system of linear equations has a solution depends heavily on the augmented matrix of the system. How to determine all the solutions (when they exist) will be the objective in Chapter 3.

EXERCISES

2.8 If, for given A and $\mathbf{b}$, the matrix equation $A\mathbf{x} = \mathbf{b}$ has more than one solution, prove that it has infinitely many solutions.

[*Hint.* If $\mathbf{x}_1$ and $\mathbf{x}_2$ are solutions, show that so also is $p\mathbf{x}_1 + q\mathbf{x}_2$ where $p + q = 1$.]

3. Equilibrium-seeking positions

Consider the following situation. In normal population movement, a certain proportion of city dwellers move into the country every year and a certain proportion of country dwellers decide to become city dwellers. A similar situation occurs in national employment where a certain percentage of unemployed people find jobs and a certain percentage of employed people become unemployed. Mathematically, these situations are essentially the same. The problem of how to describe them in a concrete mathematical way, and to answer the obvious question of whether or not such changes can ever reach a 'steady state' is one that we shall now illustrate.

To be more specific, let us suppose that 75% of the unemployed at the beginning of a year find jobs during that year, and that 5% of people with jobs become unemployed during the year. These proportions are of course somewhat optimistic, and might lead one to conjecture that 'sooner or later' everyone will have a job. But these figures are chosen to illustrate the point that we want to make, namely that the system 'settles down' to fixed proportions.

The system can be described compactly by the following matrix and its obvious interpretation:

unemployed   employed

into unemployment $\frac{1}{4}$ $\frac{1}{20}$

into employment $\frac{3}{4}$ $\frac{19}{20}$

Suppose now that the fraction of the population that is originally unemployed is L_0 and that the fraction that is originally employed is $M_0 = 1 - L_0$. We represent this state of affairs by the matrix

$$\begin{bmatrix} L_0 \\ M_0 \end{bmatrix}.$$

More generally, we let the matrix

$$\begin{bmatrix} L_i \\ M_i \end{bmatrix}$$

signify the proportions of the unemployed/employed population at the end of the i-th year. At the end of the first year we therefore have

$$L_1 = \tfrac{1}{4}L_0 + \tfrac{1}{20}M_0$$
$$M_1 = \tfrac{3}{4}L_0 + \tfrac{19}{20}M_0$$

and we can express these equations in the matrix form

$$\begin{bmatrix} L_1 \\ M_1 \end{bmatrix} = \begin{bmatrix} \tfrac{1}{4} & \tfrac{1}{20} \\ \tfrac{3}{4} & \tfrac{19}{20} \end{bmatrix} \begin{bmatrix} L_0 \\ M_0 \end{bmatrix}$$

which involves the 2×2 matrix introduced above.

Similarly, at the end of the second year we have

$$L_2 = \tfrac{1}{4}L_1 + \tfrac{1}{20}M_1$$
$$M_2 = \tfrac{3}{4}L_1 + \tfrac{19}{20}M_1$$

and consequently

$$\begin{bmatrix} L_2 \\ M_2 \end{bmatrix} = \begin{bmatrix} \tfrac{1}{4} & \tfrac{1}{20} \\ \tfrac{3}{4} & \tfrac{19}{20} \end{bmatrix} \begin{bmatrix} L_1 \\ M_1 \end{bmatrix} = \begin{bmatrix} \tfrac{1}{4} & \tfrac{1}{20} \\ \tfrac{3}{4} & \tfrac{19}{20} \end{bmatrix}^2 \begin{bmatrix} L_0 \\ M_0 \end{bmatrix}.$$

Using induction, we can say that at the end of the k-th year the relationship between L_k, M_k and L_0, M_0 is given by

$$\begin{bmatrix} L_k \\ M_k \end{bmatrix} = \begin{bmatrix} \tfrac{1}{4} & \tfrac{1}{20} \\ \tfrac{3}{4} & \tfrac{19}{20} \end{bmatrix}^k \begin{bmatrix} L_0 \\ M_0 \end{bmatrix}.$$

Now it can be shown (and we shall be able to do so much later) that, for all positive integers k, we have

$$\begin{bmatrix} \tfrac{1}{4} & \tfrac{1}{20} \\ \tfrac{3}{4} & \tfrac{19}{20} \end{bmatrix}^k = \frac{1}{16} \begin{bmatrix} 1 + \tfrac{15}{5^k} & 1 - \tfrac{1}{5^k} \\ 15(1 - \tfrac{1}{5^k}) & 15 + \tfrac{1}{5^k} \end{bmatrix}.$$

This is rather like pulling a rabbit out of a hat, for we are far from having the machinery at our disposal to obtain this result; but the reader will at least be able to verify it by induction. From this formula we can see that, the larger k becomes, the closer is the approximation

$$\begin{bmatrix} \tfrac{1}{4} & \tfrac{1}{20} \\ \tfrac{3}{4} & \tfrac{19}{20} \end{bmatrix}^k \sim \begin{bmatrix} \tfrac{1}{16} & \tfrac{1}{16} \\ \tfrac{15}{16} & \tfrac{15}{16} \end{bmatrix}.$$

Since $L_0 + M_0 = 1$, we therefore have

$$\begin{bmatrix} L_k \\ M_k \end{bmatrix} \sim \begin{bmatrix} \tfrac{1}{16} & \tfrac{1}{16} \\ \tfrac{15}{16} & \tfrac{15}{16} \end{bmatrix} \begin{bmatrix} L_0 \\ M_0 \end{bmatrix} = \begin{bmatrix} \tfrac{1}{16} \\ \tfrac{15}{16} \end{bmatrix}.$$

Put another way, irrespective of the initial values of L_0 and M_0, we see that the system is 'equilibrium-seeking' in the sense that 'eventually' one sixteenth of the population remains unemployed. Of course, the lack of any notion of a limit for a sequence of matrices precludes any rigorous description of what is meant by an 'equilibrium-seeking' system. However, only the reader's intuition is called on to appreciate this particular application.

4. Difference equations

The system of pairs of equations

$$x_{n+1} = ax_n + by_n$$
$$y_{n+1} = cx_n + dy_n$$

is called a system of **linear difference equations**. Associated with such a system
are the sequences $(x_n)_{n \geqslant 1}$ and $(y_n)_{n \geqslant 1}$, and the problem is to determine the general
values of x_n, y_n given the initial values of x_1, y_1.

The above system can be written in the matrix form $\mathbf{z}_{n+1} = A\mathbf{z}_n$ where

$$\mathbf{z}_n = \begin{bmatrix} x_n \\ y_n \end{bmatrix}, \quad A = \begin{bmatrix} a & b \\ c & d \end{bmatrix}.$$

Clearly, we have $\mathbf{z}_2 = A\mathbf{z}_1, \mathbf{z}_3 = A\mathbf{z}_2 = A^2\mathbf{z}_1$, and inductively we see that

$$\mathbf{z}_{n+1} = A^n\mathbf{z}_1.$$

Thus a solution can be found if an expression for A^n is known.

The problem of determining the high powers of a matrix is one that will also be
dealt with later.

5. A definition of complex numbers

Complex numbers are usually introduced at an elementary level by saying that
a complex number is 'a number of the form $x + iy$ where x, y are real numbers and
$i^2 = -1$'. Complex numbers add and multiply as follows:

$$(x + iy) + (x' + iy') = (x + x') + i(y + y');$$

$$(x + iy)(x' + iy') = (xx' - yy') + i(xy' + x'y).$$

Also, for every real number λ we have

$$\lambda(x + iy) = \lambda x + i\lambda y.$$

This will be familiar to the reader, even though (s)he may have little idea as to
what this number system is! For example, $i = \sqrt{-1}$ is not a real number, so what
does the product iy mean? Is $i0 = 0$? If so then every real number x can be written
as $x = x + i0$, which is familiar. This heuristic approach to complex numbers can
be confusing. However, there is a simple approach that uses 2×2 matrices which
is more illuminating and which we shall now describe. Of course, at this level we
have to contend with the fact that the reader will be equally unsure about what a real
number is, but let us proceed on the understanding that the real number system is
that to which the reader has been accustomed throughout her/his schooldays.

The essential idea behind complex numbers is to develop an algebraic system of
objects (called complex numbers) that is 'larger' than the real number system, in the
sense that it contains a replica of this system, and in which the equation $x^2 + 1 = 0$
has a solution. This equation is, of course, insoluble in the real number system.

There are several ways of 'extending' the real number system in this way, and the
one we shall describe uses 2×2 matrices. For this purpose, consider the collection
C_2 of all 2×2 matrices of the form

$$M(a, b) = \begin{bmatrix} a & b \\ -b & a \end{bmatrix},$$

where a and b are real numbers. Invoking Theorem 1.10, we can write $M(a,b)$ uniquely as the sum of a symmetric matrix and a skew-symmetric matrix:

$$\begin{bmatrix} a & b \\ -b & a \end{bmatrix} = \begin{bmatrix} a & 0 \\ 0 & a \end{bmatrix} + \begin{bmatrix} 0 & b \\ -b & 0 \end{bmatrix}.$$

Thus, if we define

$$J_2 = \begin{bmatrix} 0 & 1 \\ -1 & 0 \end{bmatrix},$$

we see that every such matrix $M(a,b)$ can be written uniquely in the form

$$M(a,b) = aI_2 + bJ_2.$$

Observe now that the collection R_2 of all 2×2 matrices in C_2 that are of the form $M(a,0) = aI_2$ is a replica of the real number system; for the matrices of this type add and multiply as follows:

$$xI_2 + yI_2 = \begin{bmatrix} x+y & 0 \\ 0 & x+y \end{bmatrix} = (x+y)I_2;$$

$$xI_2 \cdot yI_2 = \begin{bmatrix} xy & 0 \\ 0 & xy \end{bmatrix} = (xy)I_2,$$

and the replication is given by associating with every real number x the matrix $xI_2 = M(x,0)$.

Moreover, the identity matrix $I_2 = 1 \cdot I_2$ belongs to R_2, and we have

$$J_2^2 = \begin{bmatrix} 0 & 1 \\ -1 & 0 \end{bmatrix}\begin{bmatrix} 0 & 1 \\ -1 & 0 \end{bmatrix} = \begin{bmatrix} -1 & 0 \\ 0 & -1 \end{bmatrix} = -I_2,$$

so that $J_2^2 + I_2 = 0$. In other words, in the system C_2 the equation $x^2 + 1 = 0$ has a solution (namely J_2).

The usual notation for complex numbers can be derived from C_2 by writing each aI_2 as simply a, writing J_2 as i, and then writing $aI_2 + bJ_2$ as $a + bi$. Since clearly, for each scalar b, we can define $J_2 b$ to be the same as bJ_2 we have that $a + bi = a + ib$.

Observe that in the system C_2 we have

$$M(a,b) + M(a',b') = \begin{bmatrix} a & b \\ -b & a \end{bmatrix} + \begin{bmatrix} a' & b' \\ -b' & a' \end{bmatrix} = M(a+a', b+b');$$

$$M(a,b)M(a',b') = \begin{bmatrix} a & b \\ -b & a \end{bmatrix}\begin{bmatrix} a' & b' \\ -b' & a' \end{bmatrix} = M(aa'-bb', ab'+ba').$$

Under the association

$$M(a,b) \leftrightarrow a + ib,$$

these equalities reflect the usual definitions of addition and multiplication in the **system $\mathbb{C}$ of complex numbers**.

 This is far from being the entire story about $\mathbb{C}$, the most remarkable feature of which is that *every* equation of the form

$$a_n X^n + a_{n-1} X^{n-1} + \cdots + a_1 X + a_0 = 0,$$

where each $a_i \in \mathbb{C}$, has a solution.

EXERCISES

2.9 Let A be a complex 2×2 matrix, i.e. the entries of A are complex numbers. Prove that if $A^2 = 0$ then A is necessarily of the form

$$\begin{bmatrix} zw & z^2 \\ -w^2 & -zw \end{bmatrix}$$

for some $z, w \in \mathbb{C}$. By considering the matrix

$$\begin{bmatrix} 0 & 0 \\ 1 & 0 \end{bmatrix},$$

show that the same is not true for real 2×2 matrices.

2.10 The **conjugate** of a complex number $z = x + iy$ is the complex number $\bar{z} = x - iy$. The **conjugate** of a complex matrix $A = [z_{ij}]_{m \times n}$ is the matrix $\bar{A} = [\bar{z}_{ij}]_{m \times n}$. Prove that $\bar{A}' = \overline{A'}$ and that, when the sums and products are defined, $\overline{A + B} = \bar{A} + \bar{B}$ and $\overline{AB} = \bar{B}\,\bar{A}$.

2.11 A square complex matrix A is **hermitian** if $\overline{A'} = A$, and **skew-hermitian** if $\overline{A'} = -A$. Prove that $A + \overline{A'}$ is hermitian and that $A - \overline{A'}$ is skew-hermitian. Prove also that every square complex matrix can be written as the sum of a hermitian matrix and a skew-hermitian matrix.

3.
Systems of Linear Equations

We shall now consider in some detail a systematic method of solving systems of linear equations. In working with such systems, there are three basic operations involved:

(1) interchanging two equations (usually for convenience);
(2) multiplying an equation by a non-zero scalar;
(3) forming a new equation by adding one equation to another.

The operation of adding a multiple of one equation to another can be achieved by a combination of (2) and (3).

We begin by considering the following three examples.

Example 3.1

To solve the system

$$
\begin{aligned}
y + 2z &= 1 \quad (1)\\
x - 2y + z &= 0 \quad (2)\\
3y - 4z &= 23 \quad (3)
\end{aligned}
$$

we multiply equation (1) by 3 and subtract the new equation from equation (3) to obtain $-10z = 20$, whence we see that $z = -2$. It then follows from equation (1) that $y = 5$, and then by equation (2) that $x = 2y - z = 12$.

Example 3.2

Consider the system

$$
\begin{aligned}
x - 2y - 4z &= 0 \quad (1)\\
-2x + 4y + 3z &= 1 \quad (2)\\
-x + 2y - z &= 1 \quad (3)
\end{aligned}
$$

If we add together equations (1) and (2), we obtain equation (3), which is therefore superfluous. Thus we have only two equations in three unknowns. What do we mean by a solution in this case?

Example 3.3

Consider the system

$$
\begin{aligned}
x + y + z + t &= 1 \quad (1) \\
x - y - z + t &= 3 \quad (2) \\
-x - y + z - t &= 1 \quad (3) \\
-3x + y - 3z - 3t &= 4 \quad (4)
\end{aligned}
$$

Adding equations (1) and (2), we obtain $x + t = 2$, whence it follows that $y + z = -1$. Adding equations (1) and (3), we obtain $z = 1$ and consequently $y = -2$. Substituting in equation (4), we obtain $-3x - 3t = 9$ so that $x + t = -3$, which is not consistent with $x + t = 2$.

This system therefore does not have a solution. Expressed in another way, given the 4×4 matrix

$$
A = \begin{bmatrix}
1 & 1 & 1 & 1 \\
1 & -1 & -1 & 1 \\
-1 & -1 & 1 & -1 \\
-3 & 1 & -3 & -3
\end{bmatrix},
$$

there are no numbers x, y, z, t that satisfy the matrix equation

$$
A \begin{bmatrix} x \\ y \\ z \\ t \end{bmatrix} = \begin{bmatrix} 1 \\ 3 \\ 1 \\ 4 \end{bmatrix}.
$$

The above three examples were chosen to provoke the question: is there a *systematic* method of tackling systems of linear equations that

(a) avoids a haphazard manipulation of the equations;

(b) yields all the solutions when they exist;

(c) makes it clear when no solution exists?

In what follows our objective will be to obtain a complete answer to this question.

We note first that in dealing with systems of linear equations the 'unknowns' play a secondary rôle. It is in fact the coefficients (which are usually integers) that are important. Indeed, each such system is completely determined by its augmented matrix. In order to work solely with this, we consider the following **elementary row operations** on this matrix:

(1) interchange two rows;

(2) multiply a row by a non-zero scalar;

(3) add one row to another.

These elementary row operations clearly correspond to the basic operations on equations listed above.

It is important to observe that *these elementary row operations do not affect the solutions (if any) of the system.* In fact, if the original system of equations has a solution then this solution is also a solution of the system obtained by applying any of the operations (1), (2), (3); and since we can in each case perform the 'inverse' operation and thereby obtain the original system, the converse is also true.

We begin by showing that the above elementary row operations have a fundamental interpretation in terms of matrix products.

Theorem 3.1

Let P be the $m \times m$ matrix that is obtained from the identity matrix I_m by permuting its rows in some way. Then for any $m \times n$ matrix A the matrix PA is the matrix obtained from A by permuting its rows in precisely the same way.

Proof

Suppose that the i-th row of P is the j-th row of I_m. Then we have

$$(k = 1, \ldots, m) \qquad p_{ik} = \delta_{ik}.$$

Consequently, for every value of k,

$$[PA]_{ik} = \sum_{t=1}^{m} p_{it} a_{tk} = \sum_{t=1}^{m} \delta_{jt} a_{tk} = a_{jk},$$

whence we see that the i-th row of PA is the j-th row of A. $\square$

Example 3.4

Consider the matrix

$$P = \begin{bmatrix} 1 & 0 & 0 & 0 \\ 0 & 0 & 1 & 0 \\ 0 & 1 & 0 & 0 \\ 0 & 0 & 0 & 1 \end{bmatrix},$$

obtained from I_4 by permuting the second and third rows. If we consider any 4×2 matrix

$$A = \begin{bmatrix} a_1 & b_1 \\ a_2 & b_2 \\ a_3 & b_3 \\ a_4 & b_4 \end{bmatrix}$$

and we compute the product

$$PA = \begin{bmatrix} 1 & 0 & 0 & 0 \\ 0 & 0 & 1 & 0 \\ 0 & 1 & 0 & 0 \\ 0 & 0 & 0 & 1 \end{bmatrix} \begin{bmatrix} a_1 & b_1 \\ a_2 & b_2 \\ a_3 & b_3 \\ a_4 & b_4 \end{bmatrix} = \begin{bmatrix} a_1 & b_1 \\ a_3 & b_3 \\ a_2 & b_2 \\ a_4 & b_4 \end{bmatrix},$$

we see that the effect of multiplying A on the left by P is to permute the second and third rows of A.

EXERCISES

3.1 Explain the effect of left multiplication by the matrix

$$P = \begin{bmatrix} 0 & 0 & 1 & 0 \\ 0 & 0 & 0 & 1 \\ 1 & 0 & 0 & 0 \\ 0 & 1 & 0 & 0 \end{bmatrix}.$$

3.2 Explain the effect of left multiplication by the matrix

$$P = \begin{bmatrix} 0 & 0 & 1 & 0 \\ 1 & 0 & 0 & 0 \\ 0 & 1 & 0 & 0 \\ 0 & 0 & 0 & 1 \end{bmatrix}.$$

Theorem 3.2

Let A be an $m \times m$ matrix and let D be the $m \times m$ diagonal matrix

$$D = \begin{bmatrix} \lambda_1 & & & \\ & \lambda_2 & & \\ & & \ddots & \\ & & & \lambda_m \end{bmatrix}.$$

Then DA is obtained from A by multiplying the i-th row of A by λ_i for $i = 1, \ldots, m$.

Proof

Clearly, we have $d_{ij} = \lambda_i \delta_{ij}$. Consequently,

$$[DA]_{ij} = \sum_{k=1}^{m} d_{ik} a_{kj} = \sum_{k=1}^{m} \lambda_i \delta_{ik} a_{kj} = \lambda_i a_{ij},$$

and so the i-th row of DA is simply λ_i times the i-th row of A.   □

Example 3.5

Consider the matrix

$$D = \begin{bmatrix} 1 & 0 & 0 & 0 \\ 0 & \alpha & 0 & 0 \\ 0 & 0 & \beta & 0 \\ 0 & 0 & 0 & 1 \end{bmatrix},$$

obtained from I_4 by multiplying the second row by α and the third row by β. If

$$A = \begin{bmatrix} a_1 & b_1 \\ a_2 & b_2 \\ a_3 & b_3 \\ a_4 & b_4 \end{bmatrix}$$

and we compute the product

$$DA = \begin{bmatrix} 1 & 0 & 0 & 0 \\ 0 & \alpha & 0 & 0 \\ 0 & 0 & \beta & 0 \\ 0 & 0 & 0 & 1 \end{bmatrix} \begin{bmatrix} a_1 & b_1 \\ a_2 & b_2 \\ a_3 & b_3 \\ a_4 & b_4 \end{bmatrix} = \begin{bmatrix} a_1 & b_1 \\ \alpha a_2 & \alpha b_2 \\ \beta a_3 & \beta b_3 \\ a_4 & b_4 \end{bmatrix}$$

we see that the effect of multiplying A on the left by D is to multiply the second row of A by α and the third row by β.

EXERCISES

3.3 Explain the effect of left multiplication by the matrix

$$P = \begin{bmatrix} 0 & 0 & \gamma & 0 \\ 0 & 0 & 0 & \delta \\ \alpha & 0 & 0 & 0 \\ 0 & \beta & 0 & 0 \end{bmatrix}.$$

Theorem 3.3

Let P be the $m \times m$ matrix that is obtained from I_m by adding λ times the s-th row to the r-th row (where r, s are fixed with $r \neq s$). Then for any $m \times n$ matrix A the matrix PA is the matrix that is obtained from A by adding λ times the s-th row of A to the r-th row of A.

Proof

Let E_{rs}^λ denote the $m \times m$ matrix that has λ in the (r, s)-th position and 0 elsewhere. Then we have

$$[E_{rs}^\lambda]_{ij} = \begin{cases} \lambda & \text{if } i = r, j = s; \\ 0 & \text{otherwise.} \end{cases}$$

Since, by definition, $P = I_m + E_{rs}^\lambda$, it follows that

$$\begin{aligned} [PA]_{ij} &= [A + E_{rs}^\lambda A]_{ij} \\ &= [A]_{ij} + \sum_{k=1}^{m} [E_{rs}^\lambda]_{ik} [A]_{kj} \\ &= \begin{cases} [A]_{ij} & \text{if } i \neq r; \\ [A]_{rj} + \lambda [A]_{sj} & \text{if } i = r. \end{cases} \end{aligned}$$

Thus we see that PA is obtained from A by adding λ times the s-th row to the r-th row. $\square$

Example 3.6

Consider the matrix

$$P = \begin{bmatrix} 1 & \lambda & 0 \\ 0 & 1 & 0 \\ 0 & 0 & 1 \end{bmatrix}$$

which is obtained from I_3 by adding λ times the second row of I_3 to the first row. If

$$A = \begin{bmatrix} a_1 & b_1 \\ a_2 & b_2 \\ a_3 & b_3 \end{bmatrix}$$

and we compute the product

$$PA = \begin{bmatrix} 1 & \lambda & 0 \\ 0 & 1 & 0 \\ 0 & 0 & 1 \end{bmatrix} \begin{bmatrix} a_1 & b_1 \\ a_2 & b_2 \\ a_3 & b_3 \end{bmatrix} = \begin{bmatrix} a_1 + \lambda a_2 & b_1 + \lambda b_2 \\ a_2 & b_2 \\ a_3 & b_3 \end{bmatrix}$$

we see that the effect of multiplying A on the left by P is to add λ times the second row of A to the first row.

EXERCISES

3.4 Explain the effect of left multiplication by the matrix

$$\begin{bmatrix} 0 & \alpha & 1 \\ 0 & \beta & 0 \\ 1 & \gamma & 0 \end{bmatrix} .$$

Definition

By an **elementary matrix** of size $n \times n$ we shall mean a matrix that is obtained from the identity matrix I_n by applying to it a single elementary row operation.

In what follows we use the 'punning notation' ρ_i to mean 'row i'.

Example 3.7

The following are examples of 3×3 elementary matrices:

$$\begin{bmatrix} 1 & 0 & 0 \\ 0 & 0 & 1 \\ 0 & 1 & 0 \end{bmatrix} (\rho_2 \leftrightarrow \rho_3); \qquad \begin{bmatrix} 1 & 0 & 0 \\ 0 & 2 & 0 \\ 0 & 0 & 1 \end{bmatrix} (2\rho_2);$$

$$\begin{bmatrix} 1 & 0 & 1 \\ 0 & 1 & 0 \\ 0 & 0 & 1 \end{bmatrix} (\rho_1 + \rho_3)$$

Definition

In a product AB we shall say that B is **pre-multiplied** by A or, equivalently, that A is **post-multiplied** by B.

The following result is now an immediate consequence of Theorems 3.1, 3.2 and 3.3:

Theorem 3.4

An elementary row operation on an $m \times n$ matrix A is achieved on pre-multiplying A by a suitable elementary matrix. The elementary matrix in question is precisely that obtained by applying the same elementary row operation to I_m. $\square$

Having observed this important point, let us return to the system of equations described in matrix form by $A\mathbf{x} = \mathbf{b}$. It is clear that when we perform a basic operation on the equations all we do is to perform an elementary row operation on the augmented matrix $A|\mathbf{b}$. It therefore follows from Theorem 3.4 that performing a basic operation on the equations is the same as changing the system $A\mathbf{x} = \mathbf{b}$ to the system $EA\mathbf{x} = E\mathbf{b}$ where E is some elementary matrix. Moreover, the system of equations that corresponds to the matrix equation $EA\mathbf{x} = E\mathbf{b}$ is equivalent to the original system in the sense that it has the same set of solutions (if any).

Proceeding in this way, we see that to every string of k basic operations there corresponds a string of elementary matrices $E_1, \ldots, E_k$ such that the the resulting system is represented by

$$E_k \cdots E_2 E_1 A\mathbf{x} = E_k \cdots E_2 E_1 \mathbf{b},$$

which is of the form $B\mathbf{x} = \mathbf{c}$ and is equivalent to the original system.

Now the whole idea of applying matrices to solve systems of linear equations is to obtain a simple systematic method of determining a convenient final matrix B so that the solutions (if any) of the system $B\mathbf{x} = \mathbf{c}$ can be found easily, such solutions being precisely the solutions of the original system $A\mathbf{x} = \mathbf{b}$.

Our objective is to develop such a method. We shall insist that the method

(1) will avoid having to write down explicitly the elementary matrices involved at each stage;

(2) will determine automatically whether or not a solution exists;

(3) will provide all the solutions.

In this connection, there are two main problems that we have to deal with, namely

(a) what form should the matrix B have?;

(b) can our method be designed to remove all equations that may be superfluous?

These requirements add up to a tall order perhaps, but we shall see in due course that the method we shall describe meets all of them.

We begin by considering the following type of matrix.

Definition

By a **row-echelon** (or **stairstep**) matrix we shall mean a matrix of the general form

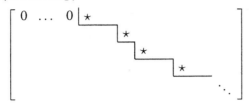

in which all the entries 'under the stairstep' are zero, all the 'corner entries' (those marked $\star$) are non-zero, and all other entries are arbitrary.

- Note that the 'stairstep' descends one row at a time and that a 'step' may traverse several columns.

Example 3.8

The 5×8 matrix

$$\begin{bmatrix} 0 & 1 & 3 & 0 & 2 & 9 & 0 & 1 \\ 0 & 0 & 0 & 4 & 0 & 3 & 5 & 1 \\ 0 & 0 & 0 & 0 & 0 & 1 & 1 & 1 \\ 0 & 0 & 0 & 0 & 0 & 0 & 0 & 0 \\ 0 & 0 & 0 & 0 & 0 & 0 & 0 & 0 \end{bmatrix}$$

is a row-echelon matrix.

Example 3.9

The 3×3 matrix

$$\begin{bmatrix} 1 & 2 & 3 \\ 0 & 4 & 5 \\ 0 & 0 & 6 \end{bmatrix}$$

is a row-echelon matrix.

Example 3.10

Every diagonal matrix in which the diagonal entries are non-zero is a row-echelon matrix.

Theorem 3.5

By means of elementary row operations, a non-zero matrix can be transformed to a row-echelon matrix.

Proof

Suppose that $A = [a_{ij}]_{m \times n}$ is a given non-zero matrix. Reading from the left, the first non-zero column of A contains at least one non-zero element. By a suitable change of rows if necessary, we can move the row containing this non-zero entry so that it becomes the first row, thereby obtaining a matrix of the form

$$B = \begin{bmatrix} 0 & \dots & 0 & b_{11} & b_{12} & \dots & b_{1k} \\ 0 & \dots & 0 & b_{21} & b_{22} & \dots & b_{2k} \\ & \vdots & & \vdots & \vdots & & \vdots \\ 0 & \dots & 0 & b_{m1} & b_{m2} & \dots & b_{mk} \end{bmatrix}$$

in which $b_{11} \neq 0$.

Now for $i = 2, 3, \dots, n$ subtract from the i-th row b_{i1}/b_{11} times the first row. This is a combination of elementary row operations and transforms B to the matrix

$$C = \begin{bmatrix} 0 & \dots & 0 & b_{11} & b_{12} & \dots & b_{1k} \\ 0 & \dots & 0 & 0 & c_{22} & \dots & c_{2k} \\ & \vdots & & \vdots & \vdots & & \vdots \\ 0 & \dots & 0 & 0 & c_{m2} & \dots & c_{mk} \end{bmatrix}$$

in which we see the beginning of the stairstep.

Now leave the first row alone and concentrate on the $(m - 1) \times (k - 1)$ matrix $[c_{ij}]$. Applying the above argument to this submatrix, we can extend the stairstep by another row. Clearly, after at most m applications of this process we arrive at a row-echelon matrix. $\square$

The above proof yields a practical method of reducing a given matrix to a row echelon matrix.

Example 3.11

$$\begin{bmatrix} 1 & 0 & 1 & 0 & 1 \\ 1 & 1 & 0 & 0 & 2 \\ 3 & 1 & 1 & 1 & 1 \\ 0 & 1 & 2 & 1 & 2 \end{bmatrix} \rightsquigarrow \begin{bmatrix} 1 & 0 & 1 & 0 & 1 \\ 0 & 1 & -1 & 0 & 1 \\ 0 & 1 & -2 & 1 & -2 \\ 0 & 1 & 2 & 1 & 2 \end{bmatrix} \begin{matrix} \\ \rho_2 - \rho_1 \\ \rho_3 - 3\rho_1 \\ {} \end{matrix}$$

$$\rightsquigarrow \begin{bmatrix} 1 & 0 & 1 & 0 & 1 \\ 0 & 1 & -1 & 0 & 1 \\ 0 & 0 & -1 & 1 & -3 \\ 0 & 0 & 3 & 1 & 1 \end{bmatrix} \begin{matrix} \\ {} \\ \rho_3 - \rho_2 \\ \rho_4 - \rho_2 \end{matrix}$$

$$\rightsquigarrow \begin{bmatrix} 1 & 0 & 1 & 0 & 1 \\ 0 & 1 & -1 & 0 & 1 \\ 0 & 0 & -1 & 1 & -3 \\ 0 & 0 & 0 & 4 & -8 \end{bmatrix} \begin{matrix} \\ {} \\ {} \\ \rho_4 + 3\rho_3 \end{matrix}$$

It should be noted carefully that the stairstep need not in general reach the bottom row, as the following examples show.

Example 3.12

$$\begin{bmatrix} 0 & 0 & 2 \\ 1 & -1 & 1 \\ -1 & 1 & -4 \end{bmatrix} \rightsquigarrow \begin{bmatrix} 1 & -1 & 1 \\ 0 & 0 & 2 \\ -1 & 1 & -4 \end{bmatrix} \quad \rho_1 \leftrightarrow \rho_2$$

$$\rightsquigarrow \begin{bmatrix} 1 & -1 & 1 \\ 0 & 0 & 2 \\ 0 & 0 & -3 \end{bmatrix} \quad \rho_3 + \rho_1$$

$$\rightsquigarrow \begin{bmatrix} 1 & -1 & 1 \\ 0 & 0 & 2 \\ 0 & 0 & 0 \end{bmatrix} \quad \rho_3 + \tfrac{3}{2}\rho_2$$

Example 3.13

$$\begin{bmatrix} 1 & -2 & 1 \\ 2 & -4 & 2 \\ -1 & 2 & -1 \end{bmatrix} \rightsquigarrow \begin{bmatrix} 1 & -2 & 1 \\ 0 & 0 & 0 \\ 0 & 0 & 0 \end{bmatrix} \quad \begin{array}{l} \rho_2 - 2\rho_1 \\ \rho_3 + \rho_1 \end{array}$$

EXERCISES

3.5 Reduce to row-echelon form the following matrices:

$$\begin{bmatrix} 1 & 2 & 3 \\ 3 & 1 & 2 \\ 5 & 5 & 8 \end{bmatrix}, \quad \begin{bmatrix} 1 & 2 & 3 \\ 3 & 1 & 2 \\ 2 & 3 & 1 \end{bmatrix}.$$

3.6 Reduce to row-echelon form the following matrix:

$$\begin{bmatrix} 1 & 2 & 0 & 3 & 1 \\ 1 & 2 & 3 & 3 & 3 \\ 1 & 0 & 1 & 1 & 3 \\ 1 & 1 & 1 & 2 & 1 \end{bmatrix}.$$

3.7 Given the matrix

$$A = \begin{bmatrix} 1 & 1 & 0 & 1 & 4 \\ 2 & 0 & 0 & 4 & 7 \\ 1 & 1 & 1 & 0 & 5 \\ 1 & -3 & -1 & -10 & \alpha \end{bmatrix},$$

determine the value of α such that, in a row-echelon form of A, the stairstep reaches the bottom.

Definition

By an **Hermite** (or **reduced row-echelon**) matrix we mean a row-echelon matrix in which every corner entry is 1 and every entry lying above a corner entry is 0.

An Hermite matrix therefore has the general form

$$
\begin{bmatrix}
0 & \dots & 0 & 1 & & 0 & & 0 & & 0 & & 0 \\
0 & \dots & 0 & 0 & \dots & 0 & 1 & & 0 & & 0 & & 0 \\
0 & \dots & 0 & 0 & \dots & 0 & 0 & \dots & 0 & 1 & & 0 & & 0 \\
0 & \dots & 0 & 0 & \dots & 0 & 0 & \dots & 0 & 0 & \dots & 0 & 1 & & 0 \\
0 & \dots & 0 & 0 & \dots & 0 & 0 & \dots & 0 & 0 & \dots & 0 & 0 & \dots & 0 & 1 \\
0 & \dots & 0 & 0 & \dots & 0 & 0 & \dots & 0 & 0 & \dots & 0 & 0 \\
& & & \dots & & & \dots & & & & \ddots
\end{bmatrix}
$$

in which the unmarked entries lying above the stairstep are arbitrary.

Example 3.14

The 4×9 matrix

$$
\begin{bmatrix}
1 & 0 & 1 & 0 & 2 & 2 & 0 & 0 & 1 \\
0 & 1 & 0 & 0 & 0 & 2 & 0 & 1 & 0 \\
0 & 0 & 0 & 0 & 0 & 0 & 1 & 2 & 1 \\
0 & 0 & 0 & 0 & 0 & 0 & 0 & 0 & 0
\end{bmatrix}
$$

is an Hermite matrix.

Example 3.15

The identity matrix I_n is an Hermite matrix.

Theorem 3.6

Every non-zero matrix A can be transformed to an Hermite matrix by means of elementary row operations.

Proof

Let Z be a row-echelon matrix obtained from A by the process described in Theorem 3.5. Divide each non-zero row of Z by the (non-zero) corner entry in that row. This makes each of the corner entries 1. Now subtract suitable multiples of every such non-zero row from every row above it to obtain an Hermite matrix. □

The systematic procedure that is described in the proof of Theorem 3.6 is best illustrated by an example.

Example 3.16

$$
\begin{bmatrix} 1 & 2 & 1 & 2 & 1 \\ 2 & 4 & 4 & 8 & 4 \\ 3 & 6 & 5 & 7 & 7 \end{bmatrix}
\rightsquigarrow
\begin{bmatrix} 1 & 2 & 1 & 2 & 1 \\ 0 & 0 & 2 & 4 & 2 \\ 0 & 0 & 2 & 1 & 4 \end{bmatrix}
\begin{matrix} \\ \rho_2 - 2\rho_1 \\ \rho_3 - 3\rho_1 \end{matrix}
$$

$$
\rightsquigarrow
\begin{bmatrix} 1 & 2 & 1 & 2 & 1 \\ 0 & 0 & 2 & 4 & 2 \\ 0 & 0 & 0 & -3 & 2 \end{bmatrix}
\begin{matrix} \\ \\ \rho_3 - \rho_2 \end{matrix}
\quad \text{row-echelon}
$$

$$
\rightsquigarrow
\begin{bmatrix} 1 & 2 & 1 & 2 & 1 \\ 0 & 0 & 1 & 2 & 1 \\ 0 & 0 & 0 & 1 & -\frac{2}{3} \end{bmatrix}
\begin{matrix} \\ \frac{1}{2}\rho_2 \\ -\frac{1}{3}\rho_3 \end{matrix}
$$

$$
\rightsquigarrow
\begin{bmatrix} 1 & 2 & 1 & 0 & \frac{7}{3} \\ 0 & 0 & 1 & 0 & \frac{7}{3} \\ 0 & 0 & 0 & 1 & -\frac{2}{3} \end{bmatrix}
\begin{matrix} \rho_1 - 2\rho_3 \\ \rho_2 - 2\rho_3 \\ \\ \end{matrix}
$$

$$
\rightsquigarrow
\begin{bmatrix} 1 & 2 & 0 & 0 & 0 \\ 0 & 0 & 1 & 0 & \frac{7}{3} \\ 0 & 0 & 0 & 1 & -\frac{2}{3} \end{bmatrix}
\begin{matrix} \rho_1 - \rho_2 \\ \\ \end{matrix}
\quad \text{Hermite}
$$

EXERCISES

3.8 Reduce to Hermite form the matrix

$$
\begin{bmatrix} 1 & -1 & 0 & -1 & -5 & -1 \\ 2 & 1 & -1 & -4 & 1 & -1 \\ 1 & 1 & 1 & -4 & -6 & 3 \\ 1 & 4 & 2 & -8 & -5 & 8 \end{bmatrix}.
$$

3.9 Reduce to Hermite form the matrices

$$
\begin{bmatrix} 0 & 0 & 0 & 1 \\ 0 & 0 & 1 & 0 \\ 0 & 1 & 0 & 0 \\ 1 & 0 & 0 & 0 \end{bmatrix},
\quad
\begin{bmatrix} 1 & 1 & 1 & 0 \\ 1 & 1 & 0 & 1 \\ 1 & 0 & 1 & 1 \\ 0 & 1 & 1 & 1 \end{bmatrix}.
$$

3.10 Reduce to Hermite form the matrices

$$
\begin{bmatrix} 1 & 0 & 0 & 1 \\ 0 & 1 & 1 & 0 \\ 0 & 1 & 1 & 0 \\ 1 & 0 & 0 & 1 \end{bmatrix},
\quad
\begin{bmatrix} 0 & 1 & 1 & 0 \\ 1 & 0 & 0 & 1 \\ 1 & 0 & 0 & 1 \\ 0 & 1 & 1 & 0 \end{bmatrix}.
$$

The astute reader will have observed that we have refrained from talking about *the* row-echelon form of a matrix. In fact, there is no unique row-echelon form. To see this, it suffices to observe that we can begin the process of reduction to row-echelon form by moving any non-zero row to become the first row. However, we *can* talk of *the* Hermite form since, as we shall see, this *is* unique. In fact, it is precisely because of this that such matrices are the focus of our attention. As far as the problem in hand is concerned, namely the solution of $A\mathbf{x} = \mathbf{b}$, we can reveal that the Hermite form of A is precisely the matrix that will satisfy the requirements we have listed above. In order to establish these facts, however, we must develop some new ideas. For this purpose, we introduce the following notation.

Given an $m \times n$ matrix $A = [a_{ij}]$ we shall use the notation

$$\mathbf{A}_i = \begin{bmatrix} a_{i1} & a_{i2} & \dots & a_{in} \end{bmatrix},$$

and quite often we shall not distinguish this from the i-th row of A. Similarly, the i-th column of A will often be confused with the column matrix

$$\mathbf{a}_i = \begin{bmatrix} a_{1i} \\ a_{2i} \\ \vdots \\ a_{mi} \end{bmatrix}.$$

Definition

By a **linear combination** of the rows (columns) of A we shall mean an expression of the form

$$\lambda_1 x_1 + \lambda_2 x_2 + \cdots + \lambda_p x_p$$

where each x_i is a row (column) of A.

Example 3.17

The row matrix $\begin{bmatrix} 2 & -3 & 1 \end{bmatrix}$ can be written in the form

$$2\begin{bmatrix} 1 & 0 & 0 \end{bmatrix} - 3\begin{bmatrix} 0 & 1 & 0 \end{bmatrix} + 1\begin{bmatrix} 0 & 0 & 1 \end{bmatrix}$$

and so is a linear combination of the rows of I_3.

Example 3.18

The column matrix

$$\begin{bmatrix} 0 \\ 4 \\ -2 \end{bmatrix}$$

can be written

$$4\begin{bmatrix} 0 \\ 1 \\ 0 \end{bmatrix} - 2\begin{bmatrix} 0 \\ 0 \\ 1 \end{bmatrix}$$

and so is a linear combination of the columns of I_3.

Example 3.19

The column matrix

$$\begin{bmatrix} 1 \\ 0 \\ 0 \end{bmatrix}$$

is not a linear combination of the columns of the matrix

$$\begin{bmatrix} 0 & 0 & 0 \\ 1 & 0 & 1 \\ 0 & 1 & 0 \end{bmatrix}.$$

Definition

If $x_1, \ldots, x_p$ are rows (columns) of A then we shall say that $x_1, \ldots, x_p$ are **linearly independent** if the only scalars $\lambda_1, \ldots, \lambda_p$ which are such that

$$\lambda_1 x_1 + \lambda_2 x_2 + \ldots + \lambda_p x_p = 0$$

are $\lambda_1 = \lambda_2 = \cdots = \lambda_p = 0$.

Expressed in an equivalent way, the rows (columns) $x_1, \ldots, x_p$ are linearly independent if the only way that the zero row (column) 0 can be expressed as a linear combination of $x_1, \ldots, x_p$ is the trivial way, namely

$$0 = 0x_1 + 0x_2 + \cdots + 0x_p.$$

If $x_1, \ldots, x_p$ are not linearly independent then they are **linearly dependent**.

Example 3.20

The columns of the identity matrix I_n are linearly independent. In fact, we have

$$\lambda_1 \begin{bmatrix} 1 \\ 0 \\ 0 \\ \vdots \\ 0 \end{bmatrix} + \lambda_2 \begin{bmatrix} 0 \\ 1 \\ 0 \\ \vdots \\ 0 \end{bmatrix} + \lambda_3 \begin{bmatrix} 0 \\ 0 \\ 1 \\ \vdots \\ 0 \end{bmatrix} + \cdots + \lambda_n \begin{bmatrix} 0 \\ 0 \\ 0 \\ \vdots \\ n \end{bmatrix} = \begin{bmatrix} \lambda_1 \\ \lambda_2 \\ \lambda_3 \\ \vdots \\ \lambda_n \end{bmatrix}$$

and this is the zero column if and only if $\lambda_1 = \lambda_2 = \cdots = \lambda_n = 0$.

Similarly, the rows of I_n are linearly independent.

Example 3.21

In the matrix

$$A = \begin{bmatrix} 1 & 2 & 0 & 2 \\ 0 & 1 & 1 & 1 \\ 1 & 0 & 1 & 0 \end{bmatrix}$$

the columns are not linearly independent. Indeed, the second and fourth columns are the same, so that $\mathbf{a}_2 = \mathbf{a}_4$ which we can write in the form $1\mathbf{a}_2 - 1\mathbf{a}_4 = \mathbf{0}$. However, the first three columns of A are linearly independent. To see this, suppose that we have $\lambda_1\mathbf{a}_1 + \lambda_2\mathbf{a}_2 + \lambda_3\mathbf{a}_3 = \mathbf{0}$. Then

$$\lambda_1\begin{bmatrix}1\\0\\1\end{bmatrix} + \lambda_2\begin{bmatrix}2\\1\\0\end{bmatrix} + \lambda_3\begin{bmatrix}0\\1\\1\end{bmatrix} = \begin{bmatrix}0\\0\\0\end{bmatrix},$$

and consequently

$$\begin{array}{rcl}
\lambda_1 + 2\lambda_2 & = & 0 \\
\lambda_2 + \lambda_3 & = & 0 \\
\lambda_1 + \lambda_3 & = & 0
\end{array}$$

from which it is easily seen that $\lambda_1 = \lambda_2 = \lambda_3 = 0$.

EXERCISES

3.11 Prove that in the matrix
$$\begin{bmatrix}1 & 2 & 0\\0 & 1 & 1\\1 & 0 & 1\end{bmatrix}$$

the rows are linearly independent.

3.12 For the matrix
$$\begin{bmatrix}1 & 2 & 0 & 3\\1 & 2 & 3 & 3\\1 & 0 & 1 & 1\\1 & 1 & 1 & 2\end{bmatrix}$$

determine the maximum number of linearly independent rows and the maximum number of linearly independent columns.

Theorem 3.7

If the rows/columns $x_1, \ldots, x_p$ are linearly independent then none can be zero.

Proof

If we had $x_i = 0$ then we could write

$$0x_1 + \cdots + 0x_{i-1} + 1x_i + 0x_{i+1} + \cdots + 0x_p = 0,$$

so that $x_1, \ldots, x_p$ would not be independent. $\square$

Theorem 3.8

The following statements are equivalent:

  (1) $x_1, \ldots, x_p$ $(p \geqslant 2)$ *are linearly dependent;*

(2) *one of the x_i can be expressed as a linear combination of the others.*

Proof

(1) $\Rightarrow$ (2) : Suppose that $x_1, \ldots, x_p$ are dependent, where $p \geqslant 2$. Then there exist $\lambda_1, \ldots, \lambda_p$ such that

$$\lambda_1 x_1 + \cdots + \lambda_p x_p = 0$$

with at least one of the λ_i not zero. Suppose that $\lambda_k \neq 0$. Then the above equation can be rewritten in the form

$$x_k = -\tfrac{\lambda_1}{\lambda_k} x_1 - \cdots - \tfrac{\lambda_p}{\lambda_k} x_p,$$

i.e. x_k is a linear combination of $x_1, \ldots, x_{k-1}, x_{k+1}, \ldots, x_p$.

(2) $\Rightarrow$ (1) : Conversely, suppose that

$$x_k = \mu_1 x_1 + \cdots + \mu_{k-1} x_{k-1} + \mu_{k+1} x_{k+1} + \cdots + \mu_p x_p.$$

Then this can be written in the form

$$\mu_1 x_1 + \cdots + \mu_{k-1} x_{k-1} + (-1) x_k + \mu_{k+1} x_{k+1} + \cdots + \mu_p x_p = 0$$

where the left-hand side is a non-trivial linear combination of $x_1, \ldots, x_p$. Thus $x_1, \ldots, x_p$ are linearly dependent. $\square$

Corollary

The rows of a matrix are linearly dependent if and only if one can be obtained from the others by means of elementary row operations.

Proof

It suffices to observe that every linear combination of the rows is, by its very definition, obtained by a sequence of elementary row operations. $\square$

Example 3.22

The rows of the matrix

$$A = \begin{bmatrix} 1 & 2 & 0 & 0 \\ 2 & 1 & -1 & 1 \\ 5 & 4 & -2 & 2 \end{bmatrix}$$

are linearly dependent. This follows from the fact that $A_3 = A_1 + 2A_2$ and from the Corollary to Theorem 3.8.

We now consider the following important notion.

Definition

By the **row rank** of a matrix we mean the maximum number of linearly independent rows in the matrix.

Example 3.23

The matrix A of the previous example is of row rank 2. To see this, recall that the three rows $\mathbf{A}_1, \mathbf{A}_2, \mathbf{A}_3$ are dependent. But the rows $\mathbf{A}_1$ and $\mathbf{A}_2$ are independent since

$$\lambda_1 \mathbf{A}_1 + \lambda_2 \mathbf{A}_2 = [\lambda_1 + 2\lambda_2 \quad 2\lambda_1 + \lambda_2 \quad -\lambda_2 \quad \lambda_2]$$

and this is zero if and only if $\lambda_1 = \lambda_2 = 0$. Hence the maximum number of independent rows is 2.

Example 3.24

The identity matrix I_n has row rank n.

Example 3.25

By Theorem 3.7, a zero matrix has row rank 0.

EXERCISES

3.13 Determine the row rank of the matrix

$$\begin{bmatrix} 1 & 2 & 5 \\ 2 & 1 & 4 \\ 0 & -1 & -2 \\ 0 & 1 & 2 \end{bmatrix}.$$

It turns out that the row rank of the augmented matrix $A|\mathbf{b}$ of the system $A\mathbf{x} = \mathbf{b}$ determines precisely how many of the equations in the system are not superfluous, so it is important to have a simple method of determining the row rank of a matrix. The next result provides the key to obtaining such a method.

Theorem 3.9

Elementary row operations do not affect row rank.

Proof

It is clear that the interchange of two rows has no effect on the maximum number of independent rows, i.e. the row rank.

If now the k-th row $\mathbf{A}_k$ is a linear combination of p other rows, which by the above we may assume to be the rows $\mathbf{A}_1, \ldots, \mathbf{A}_p$, then clearly so is $\lambda \mathbf{A}_k$ for every non-zero scalar λ. It follows by Theorem 3.8 that multiplying a row by a non-zero scalar has no effect on the row rank.

Finally, suppose that we add the i-th row to the j-th row to obtain a new j-th row, say $\mathbf{A}_j^\star = \mathbf{A}_i + \mathbf{A}_j$. Since then

$$\lambda_1 \mathbf{A}_1 + \cdots + \lambda_i \mathbf{A}_i + \cdots + \lambda_j \mathbf{A}_j^\star + \cdots + \lambda_p \mathbf{A}_p$$

$$= \lambda_1 \mathbf{A}_1 + \cdots + (\lambda_i + \lambda_j)\mathbf{A}_j + \cdots + \lambda_j \mathbf{A}_j + \cdots + \lambda_p \mathbf{A}_p,$$

it is clear that if $\mathbf{A}_1, \ldots, \mathbf{A}_i, \ldots, \mathbf{A}_j, \ldots, \mathbf{A}_p$ are linearly independent then so also are $\mathbf{A}_1, \ldots, \mathbf{A}_i, \ldots, \mathbf{A}_j^{\star}, \ldots, \mathbf{A}_p$. Thus the addition of one row to another has no effect on row rank. $\square$

Definition

A matrix B is said to be **row-equivalent** to a matrix A if B can be obtained from A by means of a finite sequence of elementary row operations.

By Theorem 3.4, we can equally assert that B is row-equivalent to A if there is a product F of elementary matrices such that $B = FA$.

Since row operations are reversible, we have that if an $m \times n$ matrix B is row-equivalent to the $m \times n$ matrix A then A is row-equivalent to B. The relation of being row-equivalent is therefore a symmetric relation on the set of $m \times n$ matrices. It is trivially reflexive; and it is transitive since if F and G are each products of elementary matrices then clearly so is FG. Thus the relation of being row equivalent is an equivalence relation on the set of $m \times n$ matrices.

The following result is an immediate consequence of Theorem 3.9.

Theorem 3.10

Row-equivalent matrices have the same row rank. $\square$

The above concepts allow us now to establish:

Theorem 3.11

Every non-zero matrix can be reduced by means of elementary row operations to a unique Hermite matrix.

Proof

By Theorem 3.6, every non-zero matrix M can be transformed by row operations to an Hermite matrix. Any two Hermite matrices obtained from M in this way are clearly row-equivalent. It suffices, therefore, to prove that if A and B are each Hermite matrices and if A and B are row-equivalent then $A = B$. This we do by induction on the number of columns.

We begin by observing that the only $m \times 1$ Hermite matrix is the column matrix

$$\begin{bmatrix} 1 \\ 0 \\ \vdots \\ 0 \end{bmatrix},$$

so the result is trivial in this case. Suppose, by way of induction, that all row-equivalent Hermite matrices of size $m \times (n - 1)$ are identical and let A and B be row-equivalent Hermite matrices of size $m \times n$. Then by Theorem 3.4 there is an $m \times n$ matrix F, which is a product of elementary matrices, such that $B = FA$.

Let $\widehat{A}$ and $\widehat{B}$ be the $m \times (n-1)$ matrices that consist of the first $n-1$ columns of A, B respectively. Then we have $\widehat{B} = F\widehat{A}$ and so $\widehat{A}$ and $\widehat{B}$ are row-equivalent. By the induction hypothesis, therefore, we have $\widehat{A} = \widehat{B}$. The result will now follow if we can show that the n-th columns of A and B are the same.

For this purpose, we observe that in an Hermite matrix every non-zero row contains a corner entry 1, and these corner entries are the only non-zero entries in their respective columns. The non-zero rows of such a matrix are therefore linearly independent, and the number of such rows (equally, the number of corner entries) is therefore the row rank of the matrix.

Now since the Hermite matrices A and B are row-equivalent, they have the same row rank and therefore the same number of corner entries. If this is r then the row rank of $\widehat{A} = \widehat{B}$ must be either r or $r-1$. In the latter case, the n-th columns of A and B consist of a corner entry 1 in the r-th row and 0 elsewhere, so these columns are equal and hence in this case $A = B$. In the former case, we deduce from $B = FA$ that, for $1 \leqslant i \leqslant r$,

$$(1) \qquad [b_{i1} \quad \dots \quad b_{in}] = \sum_{k=1}^{r} \lambda_k [a_{k1} \quad \dots \quad a_{kn}].$$

In particular, for the matrix $\widehat{A} = \widehat{B}$ we have

$$[a_{i1} \quad \dots \quad a_{i,n-1}] = [b_{i1} \quad \dots \quad b_{i,n-1}] = \sum_{k=1}^{r} \lambda_k [a_{k1} \quad \dots \quad a_{k,n-1}].$$

But since the first r rows of $\widehat{A}$ are independent we deduce from this that $\lambda_i = 1$ and $\lambda_k = 0$ for $k \neq i$. It now follows from (1) that

$$[b_{i1} \quad \dots \quad b_{in}] = [a_{i1} \quad \dots \quad a_{in}]$$

and hence that $b_{in} = a_{in}$. Thus the n-th columns of A and B coincide and so $A = B$ also in this case. $\square$

Corollary

The row rank of a matrix is the number of non-zero rows in any row-echelon form of the matrix.

Proof

Let B be a row-echelon form of A and let H be the Hermite form obtained from B. Since H is unique, the number of non-zero rows of B is precisely the number of non-zero rows of H, which is the row rank of A. $\square$

The uniqueness of the Hermite form means that two given matrices are row-equivalent if and only if they have the same Hermite form. The Hermite form of A is therefore a particularly important 'representative' in the equivalence class of A relative to the relation of being row-equivalent.

EXERCISES

3.14 Consider the matrix

$$A = \begin{bmatrix} 1 & 1 & 1 & 1 & 4 \\ 1 & \lambda & 1 & 1 & 4 \\ 1 & 1 & \lambda & 3-\lambda & 6 \\ 2 & 2 & 2 & \lambda & 6 \end{bmatrix}.$$

Show that if $\lambda \neq 1, 2$ then the row rank of A is 4. What is the row rank of A when $\lambda = 1$, and when $\lambda = 2$?

3.15 Determine whether or not the matrices

$$\begin{bmatrix} 2 & 2 & 1 \\ 1 & 3 & 1 \\ 1 & 2 & 2 \end{bmatrix}, \qquad \begin{bmatrix} 2 & 1 & -1 \\ 0 & 2 & -1 \\ -3 & -2 & 3 \end{bmatrix}$$

are row-equivalent.

Similar to the concept of an elementary row operation is that of an *elementary column operation*. To obtain this we simply replace 'row' by 'column' in the definition.

It should be noted immediately that such column operations cannot be used in the same way as row operations to solve systems of linear equations since they do not produce an equivalent system.

However, there are results concerning column operations that are 'dual' to those concerning row operations. This is because column operations on a matrix can be regarded as row operations on the transpose of the matrix. For example, from the column analogues of Theorems 3.1, 3.2 and 3.3 (proved by transposition) we have the analogue of Theorem 3.4:

Theorem 3.12

An elementary column operation on an $m \times n$ matrix can be achieved by post-multiplication by a suitable elementary matrix, namely that obtained from I_n by applying to I_n precisely the same column operation. $\square$

The notion of *column-equivalence* is dual to that of row-equivalence.

Definition

The **column rank** of a matrix is defined to be the maximum number of linearly independent columns in the matrix.

The dual of Theorem 3.9 holds, namely:

Theorem 3.13

Column operations do not affect column rank. $\square$

Since it is clear that column operations can have no effect on the independence of rows, it follows that column operations have no effect on row rank. We can therefore assert:

Theorem 3.14

Row and column rank are invariant under both row and column operations.   □

EXERCISES

3.16 Determine the row and column ranks of the matrix

$$\begin{bmatrix} 0 & 2 & 3 & -4 & 1 \\ 0 & 0 & 2 & 3 & 4 \\ 2 & 2 & -5 & 2 & 4 \\ 2 & 0 & -6 & 9 & 7 \end{bmatrix}.$$

We now ask if there is any connection between the row rank and the column rank of a matrix; i.e. if the maximum number of linearly independent rows is connected in any way with the maximum number of linearly independent columns. The answer is perhaps surprising, and bears out what the reader should have observed in the previous exercise.

Theorem 3.15

Row rank and column rank are the same.

Proof

Given a non-zero $m \times n$ matrix A, let $H(A)$ be its Hermite form. Since $H(A)$ is obtained from A by row operations it has the same row rank, p say, as A. Also, we can apply column operations to $H(A)$ without changing this row rank. Also, both A and $H(A)$ have the same column rank, since row operations do not affect column rank.

Now by suitable rearrangement of its columns $H(A)$ can be transformed into the the general form

$$\begin{bmatrix} I_p & ? \\ 0_{m-p,p} & 0_{m-p,n-p} \end{bmatrix},$$

in which the submatrix marked ? is unknown but can be reduced to 0 by further column operations using the first p columns. Thus $H(A)$ can be transformed by column operations into the matrix

$$\begin{bmatrix} I_p & 0 \\ 0 & 0 \end{bmatrix}.$$

Now by its construction this matrix has the same row rank and the same column rank as A. But clearly the row rank and the column rank of this matrix are each p. It therefore follows that the row rank and the column rank of A are the same.   □

Because of Theorem 3.15 we shall talk simply of the *rank* of a matrix, meaning by this the row rank or the column rank, whichever is appropriate. The following is now immediate:

Corollary

rank A = rank A'. $\square$

If we reflect on the proof of Theorem 3.15 we see that every non-zero $m \times n$ matrix A can be reduced by means of elementary row and column operations to a matrix of the form

$$\begin{bmatrix} I_p & 0 \\ 0 & 0 \end{bmatrix},$$

the integer p being the rank of A.

Invoking Theorems 3.4 and 3.12, we can therefore assert that there is an $m \times m$ matrix P and an $n \times n$ matrix Q, each of which is a product of elementary matrices, such that

$$PAQ = \begin{bmatrix} I_p & 0 \\ 0 & 0 \end{bmatrix}.$$

The matrix on the right-hand side of this equation is often called the **normal form** of A.

How can we find matrices P, Q such that PAQ is in normal form without having to write down at each stage the elementary row and column matrices involved?

If A is of size $m \times n$, an expedient way is to work with the array

$$\begin{array}{c|c} I_n & \\ \hline A & I_m \end{array}$$

as follows. In reducing A, we apply each row operation to the bottom m rows of this array, and each column operation to the first n columns of this array. The general stage will be an array of the form

$$\begin{array}{c|c} Y & \\ \hline XAY & X \end{array} \equiv \begin{array}{c|c} F_1 F_2 \cdots F_s & \\ \hline E_t \cdots E_2 E_1 A F_1 F_2 \cdots F_s & E_t \cdots E_2 E_1 \end{array}$$

where $E_1, E_2, \ldots, E_t$ are elementary matrices corresponding to the row operations and $F_1, F_2, \ldots, F_s$ are elementary matrices corresponding to the column operations. If N is the normal form of A, the final configuration is

$$\begin{array}{c|c} Q & \\ \hline N & P \end{array}.$$

Example 3.26

Consider the matrix

$$A = \begin{bmatrix} 1 & 2 & -1 & -2 \\ -1 & -1 & 1 & 1 \\ 0 & 1 & 2 & 1 \end{bmatrix}.$$

Applying row and column operations as described above, we obtain

$$
\left[\begin{array}{cccc|ccc}
1 & 0 & 0 & 0 & & & \\
0 & 1 & 0 & 0 & & & \\
0 & 0 & 1 & 0 & & & \\
0 & 0 & 0 & 1 & & & \\
\hline
1 & 2 & -1 & -2 & 1 & 0 & 0 \\
-1 & -1 & 1 & 1 & 0 & 1 & 0 \\
0 & 1 & 2 & 1 & 0 & 0 & 1
\end{array}\right]
\rightsquigarrow
\left[\begin{array}{cccc|ccc}
1 & 0 & 0 & 0 & & & \\
0 & 1 & 0 & 0 & & & \\
0 & 0 & 1 & 0 & & & \\
0 & 0 & 0 & 1 & & & \\
\hline
1 & 2 & -1 & -2 & 1 & 0 & 0 \\
0 & 1 & 0 & -1 & 1 & 1 & 0 \\
0 & 1 & 2 & 1 & 0 & 0 & 1
\end{array}\right]
$$

$$
\rightsquigarrow
\left[\begin{array}{cccc|ccc}
1 & 0 & 0 & 0 & & & \\
0 & 1 & 0 & 0 & & & \\
0 & 0 & 1 & 0 & & & \\
0 & 0 & 0 & 1 & & & \\
\hline
1 & 2 & -1 & -2 & 1 & 0 & 0 \\
0 & 1 & 0 & -1 & 1 & 1 & 0 \\
0 & 0 & 2 & 2 & -1 & -1 & 1
\end{array}\right]
$$

$$
\rightsquigarrow
\left[\begin{array}{cccc|ccc}
1 & 0 & 0 & 0 & & & \\
0 & 1 & 0 & 0 & & & \\
0 & 0 & 1 & 0 & & & \\
0 & 0 & 0 & 1 & & & \\
\hline
1 & 2 & -1 & -2 & 1 & 0 & 0 \\
0 & 1 & 0 & -1 & 1 & 1 & 0 \\
0 & 0 & 1 & 1 & -\frac{1}{2} & -\frac{1}{2} & \frac{1}{2}
\end{array}\right]
$$

$$
\rightsquigarrow
\left[\begin{array}{cccc|ccc}
1 & -2 & 1 & 2 & & & \\
0 & 1 & 0 & 0 & & & \\
0 & 0 & 1 & 0 & & & \\
0 & 0 & 0 & 1 & & & \\
\hline
1 & 0 & 0 & 0 & 1 & 0 & 0 \\
0 & 1 & 0 & -1 & 1 & 1 & 0 \\
0 & 0 & 1 & 1 & -\frac{1}{2} & -\frac{1}{2} & \frac{1}{2}
\end{array}\right]
$$

$$
\rightsquigarrow
\left[\begin{array}{cccc|ccc}
1 & -2 & 1 & 0 & & & \\
0 & 1 & 0 & 1 & & & \\
0 & 0 & 1 & 0 & & & \\
0 & 0 & 0 & 1 & & & \\
\hline
1 & 0 & 0 & 0 & 1 & 0 & 0 \\
0 & 1 & 0 & 0 & 1 & 1 & 0 \\
0 & 0 & 1 & 1 & -\frac{1}{2} & -\frac{1}{2} & \frac{1}{2}
\end{array}\right]
$$

$$
\rightsquigarrow
\left[\begin{array}{cccc|ccc}
1 & -2 & 1 & -1 & & & \\
0 & 1 & 0 & 1 & & & \\
0 & 0 & 1 & -1 & & & \\
0 & 0 & 0 & 1 & & & \\
\hline
1 & 0 & 0 & 0 & 1 & 0 & 0 \\
0 & 1 & 0 & 0 & 1 & 1 & 0 \\
0 & 0 & 1 & 0 & -\frac{1}{2} & -\frac{1}{2} & \frac{1}{2}
\end{array}\right].
$$

Thus we see that

$$N = \begin{bmatrix} 1 & 0 & 0 & 0 \\ 0 & 1 & 0 & 0 \\ 0 & 0 & 1 & 0 \end{bmatrix}$$

so that A is of rank 3; and

$$P = \begin{bmatrix} 1 & 0 & 0 \\ 1 & 1 & 0 \\ -\frac{1}{2} & -\frac{1}{2} & \frac{1}{2} \end{bmatrix}, \qquad Q = \begin{bmatrix} 1 & -2 & 1 & -1 \\ 0 & 1 & 0 & 1 \\ 0 & 0 & 1 & -1 \\ 0 & 0 & 0 & 1 \end{bmatrix}.$$

A simple computation on the reader's part will verify that $PAQ = N$.

- It should be noted that the matrices P and Q which render $PAQ = N$ are not in general uniquely determined. The reader should do the above example in a different way (e.g. reduce A to Hermite form before doing column operations) to obtain different P and Q.

EXERCISES

3.17 Show that the matrix

$$A = \begin{bmatrix} 1 & 2 & -3 \\ 1 & -2 & 1 \\ 5 & -2 & -3 \end{bmatrix}$$

is of rank 2 and find matrices P, Q such that

$$PAQ = \begin{bmatrix} I_2 & 0 \\ 0 & 0 \end{bmatrix}.$$

3.18 Write down all the normal forms that are possible for non-zero 4×5 matrices.

3.19 Suppose that A is an $n \times n$ matrix and that m rows of A are selected to form an $m \times n$ submatrix B. By considering the number of zero rows in the normal form, prove that rank $B \geqslant m - n +$ rank A.

Definition

We say that two $m \times n$ matrices are **equivalent** if they have the same normal form.

Since the rank of matrix is the number of non-zero rows in its normal form, it is clear from the above that *two $m \times n$ matrices are equivalent if and only if they have the same rank*.

The reader can easily check that the relation of being equivalent is an equivalence relation on the set of $m \times n$ matrices. The normal form of A is a particularly simple representative of the equivalence class of A.

EXERCISES

3.20 Prove that if A and B are row-equivalent then they are equivalent.

3.21 Prove that every square matrix is equivalent to its transpose.

3.22 Show that the following matrices are equivalent :
$$\begin{bmatrix} 1 & 2 & 1 & 5 & 3 \\ -1 & 0 & 2 & -7 & -10 \\ 1 & 2 & 4 & -1 & -6 \end{bmatrix}, \quad \begin{bmatrix} 0 & 0 & 1 & -2 & -3 \\ 0 & 1 & 0 & 2 & 1 \\ 1 & 0 & 0 & 3 & 4 \end{bmatrix}.$$

We now have to hand enough machinery to solve the problem in hand. This is dealt with in the next three results.

Theorem 3.16

If A is an $m \times n$ matrix then the homogeneous system of equations $A\mathbf{x} = \mathbf{0}$ has a non-trivial solution if and only if rank $A < n$.

Proof

Let $\mathbf{a}_i$ be the i-th column of A. Then there is a non-zero column matrix

$$\mathbf{x} = \begin{bmatrix} x_1 \\ x_2 \\ \vdots \\ x_n \end{bmatrix}$$

such that $A\mathbf{x} = \mathbf{0}$ if and only if

$$x_1\mathbf{a}_1 + x_2\mathbf{a}_2 + \cdots + x_n\mathbf{a}_n = \mathbf{0};$$

for, as is readily seen, the left-hand side of this equation is simply $A\mathbf{x}$. Hence a non-trivial (=non-zero) solution $\mathbf{x}$ exists if and only if the columns of A are linearly dependent. Since A has n columns in all, this is the case if and only if the (column) rank of A is less than n. $\square$

Theorem 3.17

A non-homogeneous system $A\mathbf{x} = \mathbf{b}$ has a solution if and only if rank A = rank $A|\mathbf{b}$.

Proof

If A is of size $m \times n$ then there is an $n \times 1$ matrix $\mathbf{x}$ such that $A\mathbf{x} = \mathbf{b}$ if and only if there are scalars $x_1, \ldots, x_n$ such that

$$x_1\mathbf{a}_1 + x_2\mathbf{a}_2 + \cdots + x_n\mathbf{a}_n = \mathbf{b}.$$

This is the case if and only if $\mathbf{b}$ is linearly dependent on the columns of A, which is the case if and only if the augmented matrix $A|\mathbf{b}$ is column-equivalent to A, i.e. has the same (column) rank as A. $\square$

Definition

We shall say that a system of linear equations is **consistent** if it has a solution (which, in the homogeneous case, is non-trivial); otherwise we shall say that it is *inconsistent*.

Theorem 3.18

Let a consistent system of linear equations have as coefficient matrix the $m \times n$ matrix A. If rank $A = p$ *then $n - p$ of the unknowns can be assigned arbitrarily and the equations can be solved in terms of them as parameters.*

Proof

Working with the augmented matrix $A|\mathbf{b}$, or simply with A in the homogeneous case, perform row operations to transform A to Hermite form. We thus obtain a matrix of the form $H(A)|\mathbf{c}$ in which, if the rank of A is p, there are p non-zero rows. The corresponding system of equations $H(A)\mathbf{x} = \mathbf{c}$ is equivalent to the original system, and its form allows us to assign $n - p$ of the unknowns as solution parameters. □

The final statement in the above proof depends on the form of $H(A)$. The assignment of unknowns as solution parameters is best illustrated by examples. This we shall now do. It should be noted that in practice there is no need to test first for consistency using Theorem 3.17 since the method of solution will determine this automatically.

Example 3.27

Let us determine for what values of α the system

$$\begin{aligned} x + y + z &= 1 \\ 2x - y + 2z &= 1 \\ x + 2y + z &= \alpha \end{aligned}$$

has a solution.

By Theorem 3.17, a solution exists if and only if the rank of the coefficient matrix of the system is the same as the rank of the augmented matrix, these ranks being determined by the number of non-zero rows in any row-echelon form.

So we begin by reducing the augmented matrix to row-echelon form:

$$\begin{bmatrix} 1 & 1 & 1 & 1 \\ 2 & -1 & 2 & 1 \\ 1 & 2 & 1 & \alpha \end{bmatrix} \leadsto \begin{bmatrix} 1 & 1 & 1 & 1 \\ 0 & -3 & 0 & -1 \\ 0 & 1 & 0 & \alpha - 1 \end{bmatrix}$$

$$\leadsto \begin{bmatrix} 1 & 1 & 1 & 1 \\ 0 & 1 & 0 & \frac{1}{3} \\ 0 & 1 & 0 & \alpha - 1 \end{bmatrix}$$

$$\leadsto \begin{bmatrix} 1 & 1 & 1 & 1 \\ 0 & 1 & 0 & \frac{1}{3} \\ 0 & 0 & 0 & \alpha - \frac{4}{3} \end{bmatrix}.$$

It is clear that the ranks are the same (and hence a solution exists) if and only if $\alpha = \frac{4}{3}$.

In this case the rank is 2 (the number of non-zero rows), and the Hermite form is

$$\begin{bmatrix} 1 & 0 & 1 & \frac{2}{3} \\ 0 & 1 & 0 & \frac{1}{3} \\ 0 & 0 & 0 & 0 \end{bmatrix}.$$

Using Theorem 3.18, we can assign $3 - 2 = 1$ of the unknowns as a solution parameter. Since the corresponding system of equations is

$$x + z = \tfrac{2}{3}$$
$$y = \tfrac{1}{3}$$

we may take as the general solution $y = \frac{1}{3}$ and $x = \frac{2}{3} - z$ where z is arbitrary.

Example 3.28

Consider now the system

$$\begin{aligned} x + y + z + \quad\quad t &= 4 \\ x + \beta y + z + \quad\quad t &= 4 \\ x + y + \beta z + (3 - \beta)t &= 6 \\ 2x + 2y + 2z \quad\quad \beta t &= 6. \end{aligned}$$

The augmented matrix of the system is

$$\begin{bmatrix} 1 & 1 & 1 & 1 & 4 \\ 1 & \beta & 1 & 1 & 4 \\ 1 & 1 & \beta & 3-\beta & 6 \\ 2 & 2 & 2 & \beta & 6 \end{bmatrix}$$

which can be reduced to row-echelon form by the operations $\rho_2 - \rho_1, \rho_3 - \rho_1$, and $\rho_4 - 2\rho_1$.

We obtain

$$(2) \qquad \begin{bmatrix} 1 & 1 & 1 & 1 & 4 \\ 0 & \beta-1 & 0 & 0 & 0 \\ 0 & 0 & \beta-1 & 2-\beta & 2 \\ 0 & 0 & 0 & \beta-2 & -2 \end{bmatrix}.$$

Now if $\beta \neq 1, 2$ then the rank of the coefficient matrix is clearly 4, as is that of the augmented matrix. By Theorem 3.18, therefore, a unique solution exists (there being no unknowns that we can assign as solution parameters).

To find the solution, we reduce the above row-echelon matrix to Hermite form:

$$\leadsto \begin{bmatrix} 1 & 1 & 1 & 1 & 4 \\ 0 & 1 & 0 & 0 & 0 \\ 0 & 0 & 1 & \frac{2-\beta}{\beta-1} & \frac{2}{\beta-1} \\ 0 & 0 & 0 & 1 & \frac{-2}{\beta-2} \end{bmatrix}$$

$$\leadsto \begin{bmatrix} 1 & 1 & 1 & 1 & 4 \\ 0 & 1 & 0 & 0 & 0 \\ 0 & 0 & 1 & 0 & 0 \\ 0 & 0 & 0 & 1 & \frac{-2}{\beta-2} \end{bmatrix}$$

$$\leadsto \begin{bmatrix} 1 & 0 & 0 & 0 & 4+\frac{2}{\beta-2} \\ 0 & 1 & 0 & 0 & 0 \\ 0 & 0 & 1 & 0 & 0 \\ 0 & 0 & 0 & 1 & \frac{-2}{\beta-2} \end{bmatrix}.$$

The system of equations that corresponds to this is

$$\begin{aligned} x &= 4 + \frac{2}{\beta-2} \\ y &= 0 \\ z &= 0 \\ t &= \frac{-2}{\beta-2} \end{aligned}$$

which gives the solution immediately.

Consider now the exceptional values. First, let $\beta = 2$. Then the matrix (2) becomes

$$\begin{bmatrix} 1 & 1 & 1 & 1 & 4 \\ 0 & 1 & 0 & 0 & 0 \\ 0 & 0 & 1 & 0 & 2 \\ 0 & 0 & 0 & 0 & -2 \end{bmatrix},$$

and in the system of equations that corresponds to this augmented matrix the final equation is

$$0x + 0y + 0z + 0t = -2,$$

which is impossible. Thus when $\beta = 2$ the system is inconsistent.

If now $\beta = 1$ then the matrix (2) becomes

$$\begin{bmatrix} 1 & 1 & 1 & 1 & 4 \\ 0 & 0 & 0 & 0 & 0 \\ 0 & 0 & 0 & 1 & 2 \\ 0 & 0 & 0 & -1 & -2 \end{bmatrix},$$

which reduces to the Hermite form

$$\begin{bmatrix} 1 & 1 & 1 & 0 & 2 \\ 0 & 0 & 0 & 1 & 2 \\ 0 & 0 & 0 & 0 & 0 \\ 0 & 0 & 0 & 0 & 0 \end{bmatrix}.$$

The corresponding system of equations is

$$x + y + z = 2,$$
$$t = 2.$$

Here the coefficient matrix is of rank 2 and so we can assign $4 - 2 = 2$ of the unknowns as solution parameters. We may therefore take as the solution

$$t = 2,$$
$$x = 2 - y - z$$

where y and z are arbitrary.

EXERCISES

3.23 Show that the system of equations

$$\begin{aligned} x + 2y + 3z + 3t &= 3 \\ x + 2y \quad\quad + 3t &= 1 \\ x \quad\quad + z + t &= 3 \\ x + y + z + 2t &= 1 \end{aligned}$$

has no solution.

3.24 For what value of α does the system of equations

$$\begin{aligned} x - 3y - z - 10t &= \alpha \\ x + y + z \quad\quad &= 5 \\ 2x \quad\quad - 4t &= 7 \\ x + y \quad + t &= 4 \end{aligned}$$

have a solution? Find the general solution when α takes this value.

3.25 Show that the equations

$$\begin{aligned} 2x + y + z &= -6\alpha \\ 2x + y + (\beta + 1)z &= 4 \\ \beta x + 3y + 2z &= 2\alpha \end{aligned}$$

has a unique solution except when $\beta = 0$ and when $\beta = 6$.

If $\beta = 0$ prove that there is only one value of α for which a solution exists, and find the general solution in this case.

Discuss the situation when $\beta = 6$.

SUPPLEMENTARY EXERCISES

3.26 Let A be a real 4×3 matrix. Determine the elementary matrices which, as post-multipliers, represent each of the following operations

(1) multiplication of the third column by -2;

(2) interchange of the second and third columns;

(3) addition of -2 times the first column to the second column.

3.27 If the real matrix

$$\begin{bmatrix} a & 1 & a & 0 & 0 & 0 \\ 0 & b & 1 & b & 0 & 0 \\ 0 & 0 & c & 1 & c & 0 \\ 0 & 0 & 0 & d & 1 & d \end{bmatrix}$$

has rank r, prove that

(1) $r > 2$;

(2) $r = 3$ if and only if $a = d = 0$ and $bc = 1$;

(3) $r = 4$ in all other cases.

3.28 Given the real matrices

$$A = \begin{bmatrix} 3 & 2 & -1 & 5 \\ 1 & -1 & 2 & 2 \\ 0 & 5 & 7 & \alpha \end{bmatrix}, \qquad B = \begin{bmatrix} 0 & 3 \\ 0 & -1 \\ 0 & 6 \end{bmatrix}$$

prove that the matrix equation $AX = B$ has a solution if and only if $\alpha = -1$.

3.29 Show that the equations

$$\begin{array}{rcrcrcrcrcl} x & - & y & & & - & u & - & 5t & = & \alpha \\ 2x & + & y & - & z & - & 4u & + & t & = & \beta \\ x & + & y & + & z & - & 4u & - & 6t & = & \gamma \\ x & + & 4y & + & 2z & - & 8u & - & 5t & = & \delta \end{array}$$

have a solution if and only if

$$8\alpha - \beta - 11\gamma + 5\delta = 0.$$

Find the general solution when $\alpha = \beta = -1, \gamma = 3, \delta = 8$.

3.30 Discuss the system of equations

$$\begin{array}{rcrcrcl} -2x & + & (\mu + 3)y & - & \lambda z & = & -3 \\ x & & & + & \lambda z & = & 1 \\ 2x & + & 4y & + & 3\lambda z & = & -\lambda. \end{array}$$

3.31 How many different normal forms are there in the set of $m \times n$ matrices?

3.32 Give an example of a matrix that cannot be reduced to normal form by means of row operations alone.

3.33 Give an example of two matrices that are equivalent but are not row-equivalent.

3.34 Prove that two matrices are equivalent if and only if one can be obtained from the other by a finite sequence of row and column operations.

3.35 Determine the values of a, b, c for which the parabola $y = ax^2 + bx + c$ passes through the three points $(-1, 1), (0, 2), (1, 3)$.

3.36 Determine the set of integer solutions of the system

$$
\begin{array}{rcrcrcrcrcl}
3x_1 &+& x_2 &+& 2x_3 &-& x_4 & & &\equiv_5& 1 \\
2x_1 &+& x_2 &+& 4x_3 &+& 3x_4 &+& 2x_5 &\equiv_5& 1 \\
x_1 &-& 2x_2 & & &+& 3x_4 &+& x_5 &\equiv_5& 3 \\
x_1 & & &-& x_3 &-& x_4 &+& 3x_5 &\equiv_5& 1 \\
2x_1 &+& 3x_2 &+& 3x_3 &-& x_4 &-& x_5 &\equiv_5& 1
\end{array}
$$

in which $\equiv_5$ denotes congruence modulo 5.

4.

Invertible Matrices

In Theorem 1.3 we showed that every $m \times n$ matrix A has an additive inverse, denoted by $-A$, which is the unique $m \times n$ matrix X that satisfies the equation $A + X = 0$. We shall now consider the multiplicative analogue of this.

Definition

Let A be an $m \times n$ matrix. Then an $n \times m$ matrix X is said to be a **left inverse** of A if it satisfies the equation $XA = I_n$; and a **right inverse** of A if it satisfies the equation $AX = I_m$.

Example 4.1

Consider the matrices

$$A = \begin{bmatrix} 1 & 1 \\ 4 & 3 \\ 3 & 4 \\ 0 & 0 \end{bmatrix}, \qquad X_{a,b} = \begin{bmatrix} -3 & 1 & 0 & a \\ -3 & 0 & 1 & b \end{bmatrix}.$$

A simple computation shows that $X_{a,b}A = I_2$, and so A *has infinitely many left inverses*. In contrast, A *has no right inverse*. To see this, it suffices to observe that if Y were a 2×4 matrix such that $AY = I_4$ then we would require $[AY]_{4,4} = 1$ which is not possible since all the entries in the fourth row of A are 0.

Example 4.2

The matrix

$$\begin{bmatrix} 1 & 0 & 0 \\ 0 & 2 & 0 \\ 0 & 0 & 3 \end{bmatrix}$$

has a common unique left inverse and unique right inverse, namely

$$\begin{bmatrix} 1 & 0 & 0 \\ 0 & \frac{1}{2} & 0 \\ 0 & 0 & \frac{1}{3} \end{bmatrix}.$$

Theorem 4.1

Let A be an m × n matrix. Then

(1) *A has a right inverse if and only if* rank $A = m$;

(2) *A has a left inverse if and only if* rank $A = n$.

Proof

(1) Suppose that the $n \times m$ matrix X is a right inverse of A, so that we have $AX = I_m$. If $\mathbf{x}_i$ denotes the i-th column of X then this equation can be expanded to the m equations

$$(i = 1, \ldots, m) \qquad A\mathbf{x}_i = \Delta_i$$

where Δ_i denotes the i-th column of I_m.

Now each of the matrix equations $A\mathbf{x}_i = \Delta_i$ represents a consistent system of m equations in n unknowns and so, by Theorem 3.17, for each i we have

$$\text{rank } A = \text{rank } A|\Delta_i.$$

Since $\Delta_1, \ldots, \Delta_m$ are linearly independent, it follows by considering column ranks that

$$
\begin{aligned}
\text{rank } A &= \text{rank } A|\Delta_1 \\
&= \text{rank } A|\Delta_1|\Delta_2 \\
&= \cdots \\
&= \text{rank } A|\Delta_1|\Delta_2|\ldots|\Delta_n = \text{rank } A|I_m = m.
\end{aligned}
$$

Conversely, suppose that the rank of A is m. Then necessarily we have that $n \geqslant m$. Consider the Hermite form of A'. Since $H(A')$ is an $n \times m$ matrix and

$$\text{rank } H(A') = \text{rank } A' = \text{rank } A = m,$$

we see that $H(A')$ is of the form

$$H(A') = \begin{bmatrix} I_m \\ 0_{n-m,m} \end{bmatrix}.$$

As this is row-equivalent to A', there exists an $n \times n$ matrix Y such that

$$YA' = \begin{bmatrix} I_m \\ 0_{n-m,m} \end{bmatrix}.$$

Taking transposes, we obtain

$$AY' = \begin{bmatrix} I_m & 0_{m,n-m} \end{bmatrix}.$$

Now let Z be the $n \times m$ matrix consisting of the first m columns of Y'. Then from the form of the immediately preceding equation we see that $AZ = I_m$, whence Z is a right inverse of A.

(2) It is an immediate consequence of Theorem 1.9 that A has a left inverse if and only if its transpose has a right inverse. The result therefore follows by applying (1) to the transpose of A. $\square$

EXERCISES

4.1 Show that the matrix

$$\begin{bmatrix} 1 & 3 & 4 & 7 \\ 2 & 3 & 5 & 8 \\ 1 & 4 & 5 & 9 \end{bmatrix}$$

has neither a left inverse nor a right inverse.

Theorem 4.2

If a matrix A has both a left inverse X and a right inverse Y then necessarily

(1) *A is square*;
(2) *X = Y*.

Proof

(1) Suppose that A is of size $m \times n$. Then by Theorem 4.1 the existence of a right inverse forces rank $A = m$, and the existence of a left inverse forces rank $A = n$. Hence $m = n$ and so A is square.

(2) If A is of size $p \times p$ then $XA = I_p = AY$ gives, by the associativity of matrix multiplication,

$$X = XI_p = X(AY) = (XA)Y = I_pY = Y. \quad \square$$

For square matrices we have the following stronger situation.

Theorem 4.3

If A is an $n \times n$ matrix then the following statements are equivalent:

(1) *A has a left inverse*;
(2) *A has a right inverse*;
(3) *A is of rank n*;
(4) *the Hermite form of A is I_n*;
(5) *A is a product of elementary matrices*.

Proof

We first establish the equivalence of (1), (2), (3), (4). That (1), (2), (3) are equivalent is immediate from Theorem 4.1.

(3) $\Rightarrow$ (4) : If A is of rank n then the Hermite form of A must have n non-zero rows, hence n corner entries 1. The only possibility is I_n.

(4) $\Rightarrow$ (3) : This is clear from the fact that rank $I_n = n$.

We complete the proof by showing that (3) $\Rightarrow$ (5) and (5) $\Rightarrow$ (3).

(3) $\Rightarrow$ (5) : If A is of rank n then, since (3) $\Rightarrow$ (1), A has a left inverse X. Since $XA = I_n$ we see that X has a right inverse A so, since (2) $\Rightarrow$ (4), there is a finite string of elementary matrices $F_1, F_2, \ldots, F_q$ such that

$$F_q \cdots F_2 F_1 X = I_n.$$

Consequently, we have

$$A = I_n A = (F_q \cdots F_2 F_1 X)A = F_q \cdots F_2 F_1 (XA) = F_q \cdots F_2 F_1$$

and so A is a product of elementary matrices.

$(5) \Rightarrow (3)$: Suppose now that $A = E_1 E_2 \cdots E_p$ where each E_i is an elementary matrix. Observe that E_p is of rank n since it is obtained from I_n by a single elementary operation which has no effect on rank. Also, pre-multiplication by an elementary matrix is equivalent to an elementary row operation, which has no effect on rank. It follows that the rank of the product $E_1 E_2 \cdots E_p$ is the same as the rank of E_p, which is n. Thus the rank of A is n. $\square$

It is immediate from the above important result that if a square matrix A has a one-sided inverse then this is a two-sided inverse (i.e. both a left inverse and a right inverse). In what follows we shall always use the word 'inverse' to mean two-sided inverse. By Theorem 4.2, inverses that exist are unique. When it exists, we denote the unique inverse of the square matrix A by A^{-1}.

Definition

If A has an inverse then we say that A is **invertible**.

If A is an invertible $n \times n$ matrix then so is A^{-1}. In fact, since $AA^{-1} = I_n = A^{-1}A$ and inverses are unique, we have that A is an inverse of A^{-1} and so $(A^{-1})^{-1} = A$.

We note at this juncture that since, by Theorem 4.3, every product of elementary matrices is invertible, we can assert that *B is row-equivalent to A if and only if there is an invertible matrix E such that B = EA*.

Another useful feature of Theorem 4.3 is that it provides a relatively simple method of determining whether or not A has an inverse, and of computing A^{-1} when it does exist. The method consists of reducing A to Hermite form: if this turns out to be I_n then A is invertible; and if the Hermite form is not I_n then A has no inverse.

In practice, just as we have seen in dealing with normal forms, there is no need to compute the elementary matrices required at each stage. We simply begin with the array $A|I_n$ and apply the elementary row operations to the entire array. In this way the process can be described by

$$A|I_n \rightsquigarrow E_1 A | E_1 \rightsquigarrow E_2 E_1 A | E_2 E_1 \rightsquigarrow \cdots .$$

At each stage we have an array of the form

$$S|Q \;\equiv\; E_i \cdots E_2 E_1 A | E_i \cdots E_2 E_1$$

in which $QA = S$. If A has an inverse then the Hermite form of A will be I_n and the final configuration will be

$$I_n | E_p \cdots E_2 E_1$$

so that $E_p \cdots E_2 E_1 A = I_n$ and consequently $A^{-1} = E_p \cdots E_2 E_1$.

Example 4.3

Consider the matrix

$$\begin{bmatrix} 1 & 2 & 3 \\ 1 & 3 & 4 \\ 1 & 4 & 4 \end{bmatrix}.$$

Applying the above procedure, we obtain

$$\begin{array}{ccc|ccc} 1 & 2 & 3 & 1 & 0 & 0 \\ 1 & 3 & 4 & 0 & 1 & 0 \\ 1 & 4 & 4 & 0 & 0 & 1 \end{array} \rightsquigarrow \begin{array}{ccc|ccc} 1 & 2 & 3 & 1 & 0 & 0 \\ 0 & 1 & 1 & -1 & 1 & 0 \\ 0 & 2 & 1 & -1 & 0 & 1 \end{array}$$

$$\rightsquigarrow \begin{array}{ccc|ccc} 1 & 2 & 3 & 1 & 0 & 0 \\ 0 & 1 & 1 & -1 & 1 & 0 \\ 0 & 0 & -1 & 1 & -2 & 1 \end{array}$$

$$\rightsquigarrow \begin{array}{ccc|ccc} 1 & 2 & 3 & 1 & 0 & 0 \\ 0 & 1 & 0 & 0 & -1 & 1 \\ 0 & 0 & -1 & 1 & -2 & 1 \end{array}$$

$$\rightsquigarrow \begin{array}{ccc|ccc} 1 & 2 & 0 & 4 & -6 & 3 \\ 0 & 1 & 0 & 0 & -1 & 1 \\ 0 & 0 & -1 & 1 & -2 & 1 \end{array}$$

$$\rightsquigarrow \begin{array}{ccc|ccc} 1 & 0 & 0 & 4 & -4 & 1 \\ 0 & 1 & 0 & 0 & -1 & 1 \\ 0 & 0 & 1 & -1 & 2 & -1 \end{array}$$

so A has an inverse, namely

$$\begin{bmatrix} 4 & -4 & 1 \\ 0 & -1 & 1 \\ -1 & 2 & -1 \end{bmatrix}.$$

EXERCISES

4.2 Determine which of the following matrices are invertible and find the inverses:

$$\begin{bmatrix} 1 & 1 & 1 \\ 1 & 2 & 3 \\ 0 & 1 & 1 \end{bmatrix}, \quad \begin{bmatrix} 1 & 2 & 1 \\ 1 & 3 & 2 \\ 1 & 0 & 1 \end{bmatrix}, \quad \begin{bmatrix} 1 & 2 & 2 \\ 1 & 3 & 1 \\ 1 & 1 & 3 \end{bmatrix},$$

$$\begin{bmatrix} 1 & 1 & 1 & 1 \\ 1 & 2 & -1 & 2 \\ 1 & -1 & 2 & 1 \\ 1 & 3 & 3 & 2 \end{bmatrix}, \quad \begin{bmatrix} 1 & 1 & 1 & 1 \\ 1 & 3 & 1 & 2 \\ 1 & 2 & -1 & 1 \\ 5 & 9 & 1 & 6 \end{bmatrix}.$$

4.3 If A is an $n \times n$ matrix prove that the homogeneous system $A\mathbf{x} = \mathbf{0}$ has only the trivial solution $\mathbf{x} = \mathbf{0}$ if and only if A is invertible.

4.4 Prove that the real 2×2 matrix

$$A = \begin{bmatrix} a & b \\ c & d \end{bmatrix}$$

is invertible if and only if $ad - bc \neq 0$, in which case find its inverse.

4.5 Determine for which values of α the matrix

$$A = \begin{bmatrix} 1 & 1 & 0 \\ 1 & 0 & 0 \\ 1 & 2 & \alpha \end{bmatrix}$$

is invertible and describe A^{-1}.

We shall see later other methods of finding inverses of square matrices. For the present, we consider some further results concerning inverses.

We first note that if A and B are invertible $n \times n$ matrices then in general $A + B$ is not invertible. This is easily illustrated by taking $A = I_n$ and $B = -I_n$ and observing that the zero $n \times n$ matrix is not invertible. However, as the following result shows, products of invertible matrices are invertible.

Theorem 4.4

Let A and B be $n \times n$ matrices. If A and B are invertible then so is AB; moreover,

$$(AB)^{-1} = B^{-1}A^{-1}.$$

Proof

It suffices to observe that

$$ABB^{-1}A^{-1} = AI_nA^{-1} = AA^{-1} = I_n$$

whence $B^{-1}A^{-1}$ is a right inverse of AB. By Theorem 4.3, AB therefore has an inverse, and $(AB)^{-1} = B^{-1}A^{-1}$. $\square$

Corollary

If A is invertible then so is A^m for every positive integer m; moreover,

$$(A^m)^{-1} = (A^{-1})^m.$$

Proof

The proof is by induction. The result is trivial for $m = 1$. As for the inductive step, suppose that it holds for m. Then, using Theorem 4.4, we have

$$(A^{-1})^{m+1} = A^{-1}(A^{-1})^m = A^{-1}(A^m)^{-1} = (A^mA)^{-1} = (A^{m+1})^{-1},$$

whence it holds for $m + 1$. $\square$

Theorem 4.5

If A is invertible then so is its transpose A'; moreover,

$$(A')^{-1} = (A^{-1})'.$$

Proof

By Theorem 4.4 we have

$$I_n = I_n' = (AA^{-1})' = (A^{-1})'A'$$

and so $(A^{-1})'$ is a left inverse, hence the inverse, of A'. □

EXERCISES

4.6 If $A_1, A_2, \ldots, A_p$ are invertible $n \times n$ matrices prove that so also is the product $A_1 A_2 \cdots A_p$, and that

$$(A_1 A_2 \cdots A_p)^{-1} = A_p^{-1} \cdots A_2^{-1} A_1^{-1}.$$

4.7 Let ϑ be a fixed real number and let

$$A = \begin{bmatrix} 0 & 1 & -\sin\vartheta \\ -1 & 0 & \cos\vartheta \\ -\sin\vartheta & \cos\vartheta & 0 \end{bmatrix}.$$

Show that $A^3 = 0$.

For each real number x define the 3×3 matrix A_x by

$$A_x = I_3 + xA + \tfrac{1}{2}x^2 A^2.$$

Prove that $A_x A_y = A_{x+y}$ and deduce that each A_x is invertible with $A_x^{-1} = A_{-x}$.

4.8 For each integer n let

$$A_n = \begin{bmatrix} 1-n & -n \\ n & 1+n \end{bmatrix}.$$

Prove that $A_n A_m = A_{n+m}$. Deduce that A_n is invertible with $A_n^{-1} = A_{-n}$.

Do the same for the matrices

$$B_n = \begin{bmatrix} 1-2n & n \\ -4n & 1+2n \end{bmatrix}.$$

What is the inverse of $A_n B_m$?

Recall that an $n \times n$ matrix is **orthogonal** if it is such that

$$AA' = I_n = A'A.$$

By Theorem 4.3, we see that in fact only one of these equalities is necessary. An orthogonal matrix is therefore an invertible matrix whose inverse is its transpose.

Example 4.4

A 2×2 matrix is orthogonal if and only if it is of one of the forms

$$\begin{bmatrix} a & b \\ -b & a \end{bmatrix}, \quad \begin{bmatrix} a & b \\ b & -a \end{bmatrix}$$

in which $a^2 + b^2 = 1$.

In fact, suppose that

$$A = \begin{bmatrix} a & b \\ c & d \end{bmatrix}$$

is orthogonal. Then from the equation

$$\begin{bmatrix} a & b \\ c & d \end{bmatrix} \begin{bmatrix} a & c \\ b & d \end{bmatrix} = AA' = I_2$$

we obtain

$$a^2 + b^2 = 1, \quad ac + bd = 0, \quad c^2 + d^2 = 1.$$

The first two equations show that if $a = 0$ then $d = 0$, and the second and third equations show that if $d = 0$ then $a = 0$.

Now when $a = d = 0$ the equations give $b = \pm 1$ and $c = \pm 1$.

On the other hand, when a and d are not zero the middle equation gives $c/d = -(b/a)$, so either $c = -b, d = a$ or $c = b, d = -a$.

It follows that in all cases the matrix is of one of the stated forms. Note that the one on the left is a rotation matrix (take $a = \cos \vartheta$ and $b = \sin \vartheta$).

Example 4.5

The matrix

$$\begin{bmatrix} 1/\sqrt{3} & 1/\sqrt{6} & -1/\sqrt{2} \\ 1/\sqrt{3} & -2/\sqrt{6} & 0 \\ 1/\sqrt{3} & 1/\sqrt{6} & 1/\sqrt{2} \end{bmatrix}$$

is orthogonal.

If P and A are $n \times n$ matrices with P invertible then for all positive integers m we have

$$(P^{-1}AP)^m = P^{-1}A^m P.$$

The proof of this is by induction. In fact the result is trivial for $m = 1$; and the inductive step follows from

$$\begin{aligned} (P^{-1}AP)^{m+1} &= P^{-1}AP(P^{-1}AP)^m \\ &= P^{-1}APP^{-1}A^m P \\ &= P^{-1}AI_n A^m P \\ &= P^{-1}A^{m+1}P. \end{aligned}$$

In certain applications it is important to be able to find an invertible matrix P such that $P^{-1}AP$ is of a particularly simple form.

Consider, for example, the case where P can be found such that $P^{-1}AP$ is a *diagonal* matrix D. Then from the above formula we have

$$D^m = (P^{-1}AP)^m = P^{-1}A^mP.$$

Consequently, by multiplying the above equation on the left by P and on the right by P^{-1}, we see that

$$A^m = PD^mP^{-1}.$$

Since D^m is easy to compute for a diagonal matrix D (simply take the m-th power of the diagonal entries), it is then an easy matter to compute A^m.

The problem of computing high powers of a matrix is one that we have seen before, in the 'equilibrium-seeking' example in Chapter 2, and this is precisely the method that is used to compute A^k in that example.

Of course, how to determine precisely when we can find an invertible matrix P such that $P^{-1}AP$ is a diagonal matrix (or some other 'nice' matrix) is quite another problem. A similar problem is that of finding under what conditions there exists an orthogonal matrix P such that $P'AP$ is a diagonal matrix.

Why we should want to be able to do this, and how to do it, are two of the most important questions in the whole of linear algebra. A full answer is very deep and has remarkable implications as far as both the theoretical and practical sides are concerned.

EXERCISES

4.9 If A is an $n \times n$ matrix such that $I_n + A$ is invertible prove that

$$(I_n - A)(I_n + A)^{-1} = (I_n + A)^{-1}(I_n - A).$$

Deduce that if

$$P = (I_n - A)(I_n + A)^{-1}$$

then P is orthogonal when A is skew-symmetric.

Given that

$$A = \begin{bmatrix} 0 & \cos\vartheta & 0 \\ -\cos\vartheta & 0 & \sin\vartheta \\ 0 & -\sin\vartheta & 0 \end{bmatrix},$$

prove that

$$P = \begin{bmatrix} \sin^2\vartheta & -\cos\vartheta & \sin\vartheta\cos\vartheta \\ \cos\vartheta & 0 & -\sin\vartheta \\ \sin\vartheta\cos\vartheta & \sin\vartheta & \cos^2\vartheta \end{bmatrix}.$$

SUPPLEMENTARY EXERCISES

4.10 Find the inverse of
$$\begin{bmatrix} 1 & 1 & 2 & 1 \\ 0 & -2 & 0 & 0 \\ 1 & 2 & 1 & -2 \\ 0 & 3 & 2 & 1 \end{bmatrix}.$$

4.11 Let A and B be row-equivalent $n \times n$ matrices. If A is invertible, prove that so also is B.

4.12 Let A and B be $n \times n$ matrices. If the product AB is invertible, prove that both A and B are invertible.

4.13 If A and B are $n \times n$ matrices with A invertible, prove that
$$(A + B)A^{-1}(A - B) = (A - B)A^{-1}(A + B).$$

4.14 If A is an $n \times n$ matrix establish the identity
$$I_n - A^{k+1} = (I_n - A)(I_n + A + A^2 + \cdots + A^k).$$

Deduce that if some power of A is the zero matrix then $I_n - A$ is invertible. Suppose now that
$$A = \begin{bmatrix} 2 & 2 & -1 & -1 \\ -1 & 0 & 0 & 0 \\ -1 & -1 & 1 & 0 \\ 0 & 1 & -1 & 1 \end{bmatrix}.$$

Compute the powers $(I_4 - A)^i$ for $i = 1, 2, 3, 4$ and, by applying the above to the matrix $A = I_4 - (I_4 - A)$, prove that A is invertible and determine A^{-1}.

4.15 Let A and B be $n \times n$ matrices such that $AB - I_n$ is invertible. Show that
$$(BA - I_n)[B(AB - I_n)^{-1}A - I_n] = I_n$$
and deduce that $BA - I_n$ is also invertible.

5.

Vector Spaces

In order to proceed further with matrices we have to take a wider view of matters. This we do through the following important notion.

Definition

By a **vector space** we shall mean a set V on which there are defined two operations, one called 'addition' and the other called 'multiplication by scalars', such that the following properties hold:

(V_1) $x + y = y + x$ for all $x, y \in V$;
(V_2) $(x + y) + z = x + (y + z)$ for all $x, y, z \in V$;
(V_3) there exists an element $0 \in V$ such that $x + 0 = x$ for every $x \in V$;
(V_4) for every $x \in V$ there exists $-x \in V$ such that $x + (-x) = 0$;
(V_5) $\lambda(x + y) = \lambda x + \lambda y$ for all $x, y \in V$ and all scalars λ;
(V_6) $(\lambda + \mu)x = \lambda x + \mu x$ for all $x \in V$ and all scalars λ, μ;
(V_7) $(\lambda\mu)x = \lambda(\mu x)$ for all $x \in V$ and all scalars λ, μ;
(V_8) $1x = x$ for all $x \in V$.

When the scalars are all real numbers we shall often talk of a **real** vector space; and when the scalars are all complex numbers we shall talk of a **complex** vector space.

- It should be noted that in the definition of a vector space the scalars need not be restricted to be real or complex numbers. They can in fact belong to any 'field' F (which may be regarded informally as a number system in which every non-zero element has a multiplicative inverse). Although in what follows we shall find it convenient to say that 'V is a vector space over a field F' to indicate that the scalars come from a field F, we shall in fact normally assume (i.e. unless explicitly mentioned otherwise) that F is either the field $\mathbb{R}$ of real numbers or the field $\mathbb{C}$ of complex numbers.

- Axioms (V_1) to (V_4) above can be summarised by saying that the algebraic structure $(V; +)$ is an **abelian group**. If we denote by F the field of scalars (usually $\mathbb{R}$ or $\mathbb{C}$) then multiplication by scalars can be considered as an **action** by F on V, described by $(\lambda, x) \mapsto \lambda x$, which relates the operations in F (addition and multiplication) to that of V (addition) in the way described by the axioms (V_5) to (V_8).

Example 5.1

Let $\text{Mat}_{m \times n} \, \mathbb{R}$ be the set of all $m \times n$ matrices with real entries. Then Theorems 1.1 to 1.4 collectively show that $\text{Mat}_{m \times n} \, \mathbb{R}$ is a real vector space under the usual operations of addition of matrices and multiplication by scalars.

Example 5.2

The set $\mathbb{R}^n$ of n-tuples $(x_1, \ldots, x_n)$ of real numbers is a real vector space under the following component-wise definitions of addition and mutiplication by scalars:

$$(x_1, \ldots, x_n) + (y_1, \ldots, y_n) = (x_1 + y_1, \ldots, x_n + y_n),$$
$$\lambda(x_1, \ldots, x_n) = (\lambda x_1, \ldots, \lambda x_n).$$

Geometrically, $\mathbb{R}^2$ represents the cartesian plane, whereas $\mathbb{R}^3$ represents three-dimensional space.

Similarly, the set $\mathbb{C}^n$ of n-tuples of complex numbers can be made into both a real vector space (with the scalars real numbers) or a complex vector space (with the scalars complex numbers).

Example 5.3

Let $\text{Map}(\mathbb{R}, \mathbb{R})$ be the set of all mappings $f : \mathbb{R} \to \mathbb{R}$. For two such mappings f, g define $f + g : \mathbb{R} \to \mathbb{R}$ to be the mapping given by the prescription

$$(f + g)(x) = f(x) + g(x),$$

and for every scalar $\lambda \in \mathbb{R}$ define $\lambda f : \mathbb{R} \to \mathbb{R}$ to be the mapping given by the prescription

$$(\lambda f)(x) = \lambda f(x).$$

Then it is readily verified that (V_1) to (V_8) are satisfied, with the rôle of the vector 0 taken by the zero mapping (i.e. the mapping ϑ such that $\vartheta(x) = 0$ for every $x \in \mathbb{R}$) and $-f$ the mapping given by $(-f)(x) = -f(x)$ for every $x \in \mathbb{R}$. These operations therefore make $\text{Map}(\mathbb{R}, \mathbb{R})$ into a real vector space.

Example 5.4

Let $\mathbb{R}_n[X]$ be the set of polynomials of degree at most n with real coefficients. The reader will recognise this as the set of objects of the form

$$a_0 + a_1 X + a_2 X^2 + \cdots + a_n X^n$$

where each $a_i \in \mathbb{R}$ and X is an 'indeterminate', the largest suffix i for which $a_i \neq 0$ being the degree of the polynomial.

We can define an addition on $\mathbb{R}_n[X]$ by setting

$$(a_0 + a_1X + \cdots + a_nX^n) + (b_0 + b_1X + \cdots + b_nX^n)$$
$$= (a_0 + b_0) + (a_1 + b_1)X + \cdots + (a_n + b_n)X^n$$

and a multiplication by scalars by

$$\lambda(a_0 + a_1X + \cdots + a_nX^n) = \lambda a_0 + \lambda a_1X + \cdots + \lambda a_nX^n.$$

In this way $\mathbb{R}_n[X]$ has the structure of a real vector space.

We now list some basic properties of the multiplication by scalars. For clarity, we shall denote (at least for a while) the additive identity element of V by 0_V and that of F (i.e. $\mathbb{R}$ or $\mathbb{C}$) by 0_F. We also use the symbol $\forall$ as an abbreviation of 'for all'.

Theorem 5.1

If V is a vector space over a field F then
   (1) $(\forall \lambda \in F) \quad \lambda 0_V = 0_V$;
   (2) $(\forall x \in V) \quad 0_F x = 0_V$;
   (3) *if $\lambda x = 0_V$ then either $\lambda = 0_F$ or $x = 0_V$;*
   (4) $(\forall x \in V)(\forall \lambda \in F) \quad (-\lambda)x = -(\lambda x) = \lambda(-x)$.

Proof

(1) By (V_3) and (V_5) we have

$$\lambda 0_V = \lambda(0_V + 0_V) = \lambda 0_V + \lambda 0_V.$$

Now add $-(\lambda 0_V)$ to each side.

   (2) By (V_6) we have

$$0_F x = (0_F + 0_F)x = 0_F x + 0_F x.$$

Now add $-(0_F x)$ to each side.

   (3) Suppose that $\lambda x = 0_V$ and that $\lambda \neq 0_F$. Then λ has a multiplicative inverse λ^{-1} and so, by (V_7) and (1), $x = 1_F x = (\lambda^{-1}\lambda)x = \lambda^{-1}(\lambda x) = \lambda^{-1}0_V = 0_V$.

   (4) By (2) and (V_6) we have

$$0_V = [\lambda + (-\lambda)]x = \lambda x + (-\lambda)x.$$

Now add $-(\lambda x)$ to each side. Also, by (1) and (V_5) we have

$$0_V = \lambda[x + (-x)] = \lambda x + \lambda(-x).$$

Now add $-(\lambda x)$ to each side. $\square$

EXERCISES

   5.1 Verify the various items of Theorem 5.1 in the particular case of the vector space $\mathrm{Mat}_{m \times n}\, \mathbb{R}$.

In order to study vector spaces we begin by concentrating on the substructures, i.e. those subsets that are themselves vector spaces.

Definition

Let V be a vector space over a field F. By a **subspace** of V we shall mean a non-empty subset W of V that is closed under the operations of V, in the sense that

(1) if $x, y \in W$ then $x + y \in W$;

(2) if $x \in W$ and $\lambda \in F$ then $\lambda x \in W$.

Note that (1) says that sums of elements of W belong also to W, and (2) says that scalar multiples of elements of W also belong to W. When these properties hold, W inherits from the parent space V all the other properties required in the definition of a vector space. For example, by taking $\lambda = -1_F$ in (2) we see that if $x \in W$ then $-x \in W$; and then, by taking $y = -x$ in (1), we obtain $0_V \in W$.

Example 5.5

Every vector space V is (trivially) a subspace of itself. V itself is therefore the biggest subspace of V.

Example 5.6

By Theorem 5.1(1), the singleton subset $\{0_V\}$ is a subspace of V. This is then the smallest subspace of V since, as observed above, we have that $0_V \in W$ for every subspace W of V.

Example 5.7

$\mathbb{R}$ is a subspace of the real vector space $\mathbb{C}$. In fact, it is clear that properties (1) and (2) above hold with $W = \mathbb{R}$ and $F = \mathbb{R}$.

Example 5.8

In the real vector space $\mathbb{R}^2$ the set $X = \{(x, 0) \; ; \; x \in \mathbb{R}\}$ is a subspace; for we have

$$(x_1, 0) + (x_2, 0) = (x_1 + x_2, 0);$$
$$\lambda(x, 0) = (\lambda x, 0),$$

and so X is closed under addition and multiplication by scalars. Thus (1) and (2) are satisfied. This subspace is simply the 'x-axis' in the cartesian plane $\mathbb{R}^2$. Similarly, the 'y-axis'

$$Y = \{(0, y) \; ; \; y \in \mathbb{R}\}$$

is a subspace of $\mathbb{R}^2$.

Example 5.9

We can expand on the previous example. In the cartesian plane $\mathbb{R}^2$ every line L through the origin has a description in the form of a particular subset of $\mathbb{R}^2$, namely

$$L = \{(x, y) \; ; \; \alpha x + \beta y = 0\}.$$

If L makes an angle ϑ with the x-axis then the gradient of L is given by $\tan \vartheta = -\alpha/\beta$.

Now if $(x_1, y_1) \in L$ and $(x_2, y_2) \in L$ then we have $\alpha x_1 = -\beta y_1$ and $\alpha x_2 = -\beta y_2$ whence $\alpha(x_1 + x_2) = -\beta(y_1 + y_2)$ and therefore

$$(x_1, y_1) + (x_2, y_2) = (x_1 + x_2, y_1 + y_2) \in L;$$

and if $(x_1, y_1) \in L$ then $\alpha x_1 = -\beta y_1$ gives $\alpha \lambda x_1 = -\beta \lambda y_1$ for every $\lambda \in \mathbb{R}$, so that

$$\lambda(x_1, y_1) = (\lambda x_1, \lambda y_1) \in L.$$

Thus we see that every line L that passes through the origin is a subspace of $\mathbb{R}^2$.

As we shall see later, apart from $\{(0,0)\}$ and $\mathbb{R}^2$ itself, these are the only subspaces of $\mathbb{R}^2$.

Example 5.10

In the cartesian space $\mathbb{R}^3$ every plane through the origin has a description in the form

$$P = \{(x, y, z) \; ; \; \alpha x + \beta y + \gamma z = 0\}.$$

To see the geometry of this, observe that if we fix z, say $z = k$, then 'the plane $z = k$' (i.e. the set $\{(x, y, k) \; ; \; x, y \in \mathbb{R}\}$) slices through P in the line

$$\{(x, y, k) \; ; \; \alpha x + \beta y = -\gamma k\}.$$

Now if $(x_1, y_1, z_1) \in P$ and $(x_2, y_2, z_2) \in P$ then it is readily seen that

$$(x_1, y_1, z_1) + (x_2, y_2, z_2) = (x_1 + x_2, y_1 + y_2, z_1 + z_2) \in P;$$

and if $(x_1, y_1, z_1) \in P$ then, for every $\lambda \in \mathbb{R}$,

$$\lambda(x_1, y_1, z_1) = (\lambda x_1, \lambda y_1, \lambda z_1) \in P.$$

Thus we see that every plane through the origin is a subspace of $\mathbb{R}^3$.

We shall see later that, apart from $\{(0,0,0)\}$ and $\mathbb{R}^3$ itself, the only subspaces of $\mathbb{R}^3$ are lines through the origin and planes through the origin.

Example 5.11

An $n \times n$ matrix over a field F is said to be **lower triangular** if it is of the form

$$\begin{bmatrix} a_{11} & 0 & 0 & \cdots & 0 \\ a_{21} & a_{22} & 0 & \cdots & 0 \\ a_{31} & a_{32} & a_{33} & \cdots & 0 \\ \vdots & \vdots & \vdots & \ddots & \vdots \\ a_{n1} & a_{n2} & a_{n3} & \cdots & a_{nn} \end{bmatrix}$$

i.e. if $a_{ij} = 0$ whenever $i < j$.

The set of lower triangular $n \times n$ matrices is a subspace of the vector space $\text{Mat}_{n \times n} F$. In fact, if A and B are lower triangular then clearly so is $A + B$ and so is λA.

EXERCISES

5.2 In the vector space $\mathbb{R}^4$ of 4-tuples of real numbers, determine which of the following subsets are subspaces:

(1) $\{(x, y, z, t) ; x = y, z = t\}$;

(2) $\{(x, y, z, t) ; x + y + z + t = 0\}$;

(3) $\{(x, y, z, t) ; x = 1\}$;

(4) $\{(x, y, z, t) ; xt = yz\}$.

5.3 Consider the set $\text{Con}(\mathbb{R}, \mathbb{R})$ of all continuous functions $f : \mathbb{R} \to \mathbb{R}$. What well-known theorems from analysis ensure that $\text{Con}(\mathbb{R}, \mathbb{R})$ is a subspace of the vector space $\text{Map}(\mathbb{R}, \mathbb{R})$?

5.4 Let $\text{Diff}(\mathbb{R}, \mathbb{R})$ be the subset of $\text{Con}(\mathbb{R}, \mathbb{R})$ consisting of all differentiable functions $f : \mathbb{R} \to \mathbb{R}$. What well-known theorems from analysis ensure that $\text{Diff}(\mathbb{R}, \mathbb{R})$ is a subspace of the vector space $\text{Con}(\mathbb{R}, \mathbb{R})$?

5.5 Determine which of the following are subspaces of the vector space $\text{Mat}_{n \times n} \mathbb{R}$:

(1) the set of symmetric $n \times n$ matrices;

(2) the set of invertible $n \times n$ matrices;

(3) the set of non-invertible $n \times n$ matrices.

5.6 If A is a real $m \times n$ matrix prove that the solutions of the homogeneous system $A\mathbf{x} = \mathbf{0}$ form a subspace of the vector space $\text{Mat}_{n \times 1} \mathbb{R}$.

5.7 Show that a line in $\mathbb{R}^3$ that does not pass through the origin cannot be a subspace of $\mathbb{R}^3$.

As the above examples illustrate, in order to show that a given set with operations is a vector space it is often very easy to do so by proving that it is a subspace of some well-known and considerably larger vector space.

Suppose now that A and B are subspaces of a vector space V over a field F and consider their intersection $A \cap B$. Since 0_V must belong to every subspace we have that $0_V \in A$ and $0_V \in B$, and therefore $0_V \in A \cap B$ so that $A \cap B \neq \emptyset$. Now if $x, y \in A \cap B$ then $x, y \in A$ gives $x + y \in A$, and $x, y \in B$ gives $x + y \in B$, whence we have that $x + y \in A \cap B$. Likewise, if $x \in A \cap B$ then $x \in A$ gives $\lambda x \in A$, and $x \in B$ gives $\lambda x \in B$, whence we have that $\lambda x \in A \cap B$. Thus we see that $A \cap B$ is also a subspace of V.

We can in fact prove a much more general statement:

Theorem 5.2

The intersection of any set of subspaces of a vector space V is a subspace of V.

Proof

Let C be a set of subspaces of V and let T be their intersection. Then $T \neq \emptyset$ since every subspace of V (and therefore every subspace in C) contains 0_V, whence so also does T.

Suppose now that $x, y \in T$. Since x and y belong to every subspace W in the set C, so does $x + y$ and hence $x + y \in T$. Also, if $x \in T$ then x belongs to every subspace W in the set C, whence so does λx and so $\lambda x \in T$. Thus we see that T is a subspace of V. $\square$

In contrast with the above situation, we note that the union of a set of subspaces of a vector space V need not be a subspace of V:

Example 5.12

In $\mathbb{R}^2$ the x-axis X and the y-axis Y are subspaces, but $X \cup Y$ is not. For example, we have $(1, 0) \in X$ and $(0, 1) \in Y$, but

$$(1, 0) + (0, 1) = (1, 1) \notin X \cup Y$$

so the subset $X \cup Y$ is not closed under addition and therefore cannot be a subspace.

Suppose now that we are given a subset S of a vector space V (with no restrictions, so that S may be empty if we wish). The collection C of all the sub*spaces* of V that contain S is not empty, for clearly V itself belongs to C. By Theorem 5.2, the intersection of all the subspaces in C is also a subspace of V, and clearly this intersection also contains S. This intersection is therefore the smallest subspace of V that contains S (and is, of course, S itself whenever S is a sub*space*). We shall denote this subspace by $\langle S \rangle$.

Example 5.13

In $\mathbb{R}^2$ consider the singleton subset $S = \{(x, y)\}$. Then $\langle S \rangle$ is the line joining (x, y) to the origin.

Our immediate objective is to characterise the subspace $\langle S \rangle$ in a useful alternative way. For this purpose, we consider first the case where S is not empty and introduce the following notion.

Definition

Let V be a vector space over a field F and let S be a non-empty subset of V. Then we say that $v \in V$ is a **linear combination of elements of** S if there exist $x_1, \ldots, x_n \in S$

and $\lambda_1, \ldots, \lambda_n \in F$ such that

$$v = \lambda_1 x_1 + \cdots + \lambda_n x_n = \sum_{i=1}^{n} \lambda_i x_i.$$

It is clear that if $v = \sum_{i=1}^{n} \lambda_i x_i$ and $w = \sum_{i=1}^{m} \mu_i y_i$ are linear combinations of elements of S then so is $v + w$; moreover, so is λv for every $\lambda \in F$. Thus the set of linear combinations of elements of S is a subspace of V. We call this the **subspace spanned by** S and denote it by Span S.

The above notions come together in the following result.

Theorem 5.3

$\langle S \rangle = $ Span S.

Proof

For every $x \in S$ we have $x = 1_F x \in$ Span S and therefore we see that $S \subseteq$ Span S. Since, by definition, $\langle S \rangle$ is the smallest subspace that contains S, and since Span S is a subspace, we see that $\langle S \rangle \subseteq$ Span S.

For the reverse inclusion, let $x_1, \ldots, x_n \in S$ and $\lambda_1, \ldots, \lambda_n \in F$. If W is any subspace of V that contains S we clearly have $x_1, \ldots, x_n \in W$ and

$$\lambda_1 x_1 + \cdots + \lambda_n x_n \in W.$$

Consequently we see that Span $S \subseteq W$. Taking W in particular to be $\langle S \rangle$, we obtain the result.   □

An important special case of the above arises when Span S is the whole of V. In this case we often say that S is a **spanning set** of V.

Example 5.14

Consider the subset $S = \{(1,0),(0,1)\}$ of the cartesian plane $\mathbb{R}^2$. For every $(x,y) \in \mathbb{R}^2$ we have

$$(x,y) = (x,0) + (0,y) = x(1,0) + y(0,1),$$

so that every element of $\mathbb{R}^2$ is a linear combination of elements of S. Thus S is a spanning set of $\mathbb{R}^2$.

Example 5.15

More generally, if the n-tuple

$$e_i = (0, \ldots, 0, 1, 0, \ldots, 0)$$

has the 1 in the i-th position then for every $(x_1, \ldots, x_n) \in \mathbb{R}^n$ we have

$$(x_1, \ldots, x_n) = x_1 e_1 + \cdots + x_n e_n.$$

Consequently, $\{e_1, \ldots, e_n\}$ spans $\mathbb{R}^n$.

Example 5.16

In $\mathbb{R}^3$ we have

$$\text{Span}\{(1,0,0)\} = \{\lambda(1,0,0) \; ; \; \lambda \in \mathbb{R}\} = \{(\lambda,0,0) \; ; \; \lambda \in \mathbb{R}\};$$

i.e. the subspace of $\mathbb{R}^3$ spanned by the singleton $\{(1,0,0)\}$ is the x-axis.

Example 5.17

In $\mathbb{R}^3$ we have

$$\text{Span}\{(1,0,0),(0,0,1)\} = \{x(1,0,0) + z(0,0,1) \; ; \; x,z \in \mathbb{R}\}$$
$$= \{(x,0,z) \; ; \; x,z \in \mathbb{R}\};$$

i.e. the subspace of $\mathbb{R}^3$ spanned by the subset $\{(1,0,0),(0,0,1)\}$ is the 'x,z-plane'.

EXERCISES

5.8 Show that for all $a,b,c \in \mathbb{R}$ the system of equations

$$
\begin{array}{rcrcrcl}
x &+& y &+& z &=& a \\
x &+& 2y &+& 3z &=& b \\
x &+& 3y &+& 2z &=& c
\end{array}
$$

is consistent. Deduce that the column matrices

$$\begin{bmatrix} 1 \\ 1 \\ 1 \end{bmatrix}, \quad \begin{bmatrix} 1 \\ 2 \\ 3 \end{bmatrix}, \quad \begin{bmatrix} 1 \\ 3 \\ 2 \end{bmatrix}$$

span the vector space $\text{Mat}_{3\times 1}\,\mathbb{R}$.

5.9 Let $\mathbb{R}_2[X]$ be the vector space of all real polynomials of degree at most 2. Consider the following elements of $\mathbb{R}_2[X]$:

$$p(X) = 1 + 2X + X^2, \quad q(X) = 2 + X^2.$$

Does $\{p(X),q(X)\}$ span $\mathbb{R}_2[X]$?

5.10 Does the set

$$\left\{ \begin{bmatrix} 1 & 1 \\ 0 & 0 \end{bmatrix}, \begin{bmatrix} 0 & 0 \\ 1 & 1 \end{bmatrix}, \begin{bmatrix} 1 & 0 \\ 0 & 1 \end{bmatrix}, \begin{bmatrix} 0 & 1 \\ 1 & 1 \end{bmatrix} \right\}$$

span the vector space $\text{Mat}_{2\times 2}\,\mathbb{R}$?

We now formalise, for an arbitrary vector space, a notion that we have seen before in dealing with matrices.

Definition

Let S be a non-empty subset of a vector space V over a field F. Then S is said to be **linearly independent** if the only way of expressing 0_V as a linear combination of elements of S is the trivial way (in which all scalars are 0_F).

Equivalently, S is linearly independent if, for any given $x_1, \ldots, x_n \in S$, we have

$$\lambda_1 x_1 + \cdots + \lambda_n x_n = 0_V \implies \lambda_1 = \cdots = \lambda_n = 0_F.$$

Example 5.18

The subset $\{(1,0),(0,1)\}$ of $\mathbb{R}^2$ is linearly independent. For, if $\lambda_1(1,0) + \lambda_2(0,1) = (0,0)$ then $(\lambda_1, \lambda_2) = (0,0)$ and hence $\lambda_1 = \lambda_2 = 0_F$.

Example 5.19

More generally, if $e_i = (0, \ldots, 0, 1, 0, \ldots, 0)$ with the 1 in the i-th position then $\{e_1, \ldots, e_n\}$ is a linearly independent subset of the vector space $\mathbb{R}^n$.

Example 5.20

Every singleton subset $\{x\}$ of a vector space V with $x \neq 0$ is linearly independent. This is immediate from Theorem 5.1(3).

The following result is proved exactly as in Theorem 3.7:

Theorem 5.4

No linearly independent subset of a vector space V can contain 0_V. $\square$

A subset that is not linearly independent is said to be **linearly dependent**.

Note that, by the last example above, every dependent subset other than $\{0_V\}$ must contain at least two elements.

Linearly dependent subsets can be characterised in the following useful way, the proof of which is exactly as in that of Theorem 3.8:

Theorem 5.5

Let V be a vector space over a field F. If S is a subset of V that contains at least two elements then the following statements are equivalent:

(1) *S is linearly dependent;*

(2) *at least one element of S can be expressed as a linear combination of the other elements of S.* $\square$

Example 5.21

The subset $\{(1,1,0),(2,5,3),(0,1,1)\}$ of $\mathbb{R}^3$ is linearly dependent. In fact, we have

$$(2,5,3) = 2(1,1,0) + 3(0,1,1).$$

Example 5.22

In the vector space $\mathbb{R}_2[X]$ consider

$$p(X) = 2 + X + X^2, \quad q(X) = X + 2X^2, \quad r(X) = 2 + 2X + 3X^2.$$

A general linear combination

$$\lambda_1 p(X) + \lambda_2 q(X) + \lambda_3 r(X)$$

of these vectors is

$$2\lambda_1 + 2\lambda_3 + (\lambda_1 + \lambda_2 + 2\lambda_3)X + (\lambda_1 + 2\lambda_2 + 3\lambda_3)X^2.$$

This is the zero polynomial if and only if each of the coefficients is 0; i.e. if and only if

$$\begin{bmatrix} 2 & 0 & 2 \\ 1 & 1 & 2 \\ 1 & 2 & 3 \end{bmatrix} \begin{bmatrix} \lambda_1 \\ \lambda_2 \\ \lambda_3 \end{bmatrix} = \begin{bmatrix} 0 \\ 0 \\ 0 \end{bmatrix}.$$

The reader can easily verify that the above coefficient matrix is of rank 2 so that, by Theorem 3.16, a non-trivial solution exists. Hence there are scalars $\lambda_1, \lambda_2, \lambda_3$ which are not all zero such that

$$\lambda_1 p(X) + \lambda_2 q(X) + \lambda_3 r(X) = 0$$

and so the given set is linearly dependent.

EXERCISES

5.11 Let S_1 and S_2 be non-empty subsets of a vector space such that $S_1 \subseteq S_2$. Prove that

(1) if S_2 is linearly independent then so is S_1;

(2) if S_1 is linearly dependent then so is S_2.

5.12 Determine which of the following subsets of $\text{Mat}_{3 \times 1} \mathbb{R}$ are linearly dependent. For those that are, express one vector as a linear combination of the others:

(1) $\left\{ \begin{bmatrix} 1 \\ 0 \\ 0 \end{bmatrix}, \begin{bmatrix} -1 \\ 2 \\ 1 \end{bmatrix}, \begin{bmatrix} 2 \\ 1 \\ 1 \end{bmatrix} \right\};$

(2) $\left\{ \begin{bmatrix} 1 \\ 0 \\ 0 \end{bmatrix}, \begin{bmatrix} 1 \\ 2 \\ -1 \end{bmatrix}, \begin{bmatrix} 0 \\ 1 \\ 1 \end{bmatrix} \right\};$

(3) $\left\{ \begin{bmatrix} 1 \\ 1 \\ -1 \end{bmatrix}, \begin{bmatrix} 0 \\ 1 \\ 1 \end{bmatrix}, \begin{bmatrix} 1 \\ 2 \\ 0 \end{bmatrix} \right\}.$

5.13 Determine which of the following subsets of $\mathrm{Mat}_{2\times 2}\,\mathbb{R}$ are linearly dependent. For those that are, express one matrix as a linear combination of the others:

(1) $\left\{ \begin{bmatrix} 1 & 1 \\ 1 & 1 \end{bmatrix}, \begin{bmatrix} 1 & 0 \\ 0 & 2 \end{bmatrix}, \begin{bmatrix} 0 & 2 \\ 0 & 2 \end{bmatrix} \right\};$

(2) $\left\{ \begin{bmatrix} 1 & 1 \\ 1 & 1 \end{bmatrix}, \begin{bmatrix} 2 & 3 \\ 1 & 2 \end{bmatrix}, \begin{bmatrix} 2 & 2 \\ 1 & 1 \end{bmatrix}, \begin{bmatrix} 3 & 1 \\ 2 & 1 \end{bmatrix} \right\};$

(3) $\left\{ \begin{bmatrix} 1 & 0 \\ 0 & 2 \end{bmatrix}, \begin{bmatrix} 1 & 1 \\ 2 & 1 \end{bmatrix}, \begin{bmatrix} 0 & 3 \\ 2 & 1 \end{bmatrix}, \begin{bmatrix} 2 & 3 \\ 4 & 3 \end{bmatrix} \right\}.$

5.14 Determine which of the following subsets of $\mathbb{R}_2[X]$ are linearly dependent. For those that are, express one vector as a linear combination of the others:

(1) $\{X, 3 + X^2, X + 2X^2\};$

(2) $\{-2 + X, 3 + X, 1 + X^2\};$

(3) $\{-5 + X + 3X^2, 13 + X, 1 + X + 2X^2\}.$

We now combine the notions of linearly independent set and spanning set to obtain the following important concept.

Definition

A **basis** of a vector space V is a linearly independent subset of V that spans V.

Example 5.23

The subset $\{(1,0),(0,1)\}$ is a basis of the cartesian plane $\mathbb{R}^2$. Likewise, the subset $\{(1,0,0),(0,1,0),(0,0,1)\}$ is a basis of $\mathbb{R}^3$. More generally, $\{e_1,\dots,e_n\}$ is a basis for $\mathbb{R}^n$ where

$$e_i = (0,\dots,0,1,0,\dots,0),$$

the 1 being in the i-th position.

These bases are called the **natural** (or **canonical**) bases.

Example 5.24

In $\mathbb{R}^2$ the subset

$$\{(1,1),(1,-1)\}$$

is a basis. In fact, for every $(x,y) \in \mathbb{R}^2$ we have

$$(x,y) = \lambda_1(1,1) + \lambda_2(1,-1)$$

where $\lambda_1 = \frac{1}{2}(x+y)$ and $\lambda_2 = \frac{1}{2}(x-y)$. Thus $\{(1,1),(1,-1)\}$ spans $\mathbb{R}^2$; and if

$$\alpha(1,1) + \beta(1,-1) = (0,0)$$

then $\alpha + \beta = 0$ and $\alpha - \beta = 0$ whence $\alpha = \beta = 0$, so $\{(1, 1), (1, -1)\}$ is also linearly independent.

EXERCISES

5.15 Prove that the monomials $1, X, \ldots, X^n$ form a basis (the **natural basis**) for $\mathbb{R}_n[X]$.

5.16 Prove that the $m \times n$ matrices E_{pq} described by

$$[E_{pq}]_{ij} = \begin{cases} 1 & \text{if } i = p, \ j = q; \\ 0 & \text{otherwise} \end{cases}$$

form a basis (the **natural basis**) for $\text{Mat}_{m \times n} \mathbb{R}$.

5.17 Prove that the diagonal $n \times n$ matrices form a subspace of $\text{Mat}_{n \times n} \mathbb{R}$ and determine a basis of it.

5.18 An $n \times n$ matrix all of whose diagonal entries are the same is called a **Toeplitz matrix**. Prove that the set of Toeplitz matrices is a subspace of $\text{Mat}_{n \times n} \mathbb{R}$ and exhibit a basis for this subspace.

5.19 Let $f, g, h : \mathbb{R} \to \mathbb{R}$ be the mappings given by

$$f(x) = \cos^2 x, \quad g(x) = \sin^2 x, \quad h(x) = \cos 2x.$$

Consider the subspace of $\text{Diff}(\mathbb{R}, \mathbb{R})$ given by $W = \text{Span}\{f, g, h\}$. Find a basis for W.

A fundamental characterisation of bases is the following.

Theorem 5.6

A non-empty subset S of a vector space V is a basis of V if and only if every element of V can be expressed in a unique way as a linear combination of elements of S.

Proof

$\Rightarrow$: Suppose first that S is a basis of V. Then $V = \text{Span } S$ and so, by Theorem 5.3, every $x \in V$ is a linear combination of elements of S. Now since S is linearly independent, only one such linear combination is possible for each $x \in V$; for if $\sum \lambda_i x_i = \sum \mu_i x_i$ where $x_i \in S$ then $\sum (\lambda_i - \mu_i) x_i = 0_V$ whence each $\lambda_i - \mu_i = 0_V$ and therefore $\lambda_i = \mu_i$ for each i.

$\Leftarrow$: Conversely, suppose that every element of V can be expressed in a unique way as a linear combination of elements of S. Then, by Theorem 5.3, Span S is the whole of V. Moreover, by the hypothesis, 0_V can be expressed in only one way as a linear combination of elements of S. This can only be the linear combination in which all the scalars are 0_F. It follows, therefore, that S is also linearly independent. Hence S is a basis of V. $\quad \square$

Example 5.25

For $i = 1, \ldots, n$ let

$$a_i = (a_{i1}, a_{i2}, \ldots, a_{in}).$$

Then the set $\{a_1, \ldots, a_n\}$ is a basis of $\mathbb{R}^n$ if and only if the matrix $A = [a_{ij}]_{n \times n}$ is invertible (or, equivalently, has maximum rank n).

To see this, let $x = (x_1, \ldots, x_n) \in \mathbb{R}^n$ and consider the equation

$$x = \lambda_1 a_1 + \lambda_2 a_2 + \cdots + \lambda_n a_n.$$

By equating corresponding components, we see that this is equivalent to the system

$$x_1 = \lambda_1 a_{11} + \lambda_2 a_{21} + \cdots + \lambda_n a_{n1}$$
$$x_2 = \lambda_1 a_{12} + \lambda_2 a_{22} + \cdots + \lambda_n a_{n2}$$
$$\vdots$$
$$x_n = \lambda_1 a_{1n} + \lambda_2 a_{2n} + \cdots + \lambda_n a_{nn}$$

i.e. to the system

$$A' \begin{bmatrix} \lambda_1 \\ \vdots \\ \lambda_n \end{bmatrix} = \begin{bmatrix} x_1 \\ \vdots \\ x_n \end{bmatrix}$$

where $A = [a_{ij}]_{n \times n}$.

From these observations we see that every $x \in \mathbb{R}^n$ can be written uniquely as a linear combination of $a_1, \ldots, a_n$ if and only if the above matrix equation has a unique solution. This is so if and only if A' is invertible, which is the case if and only if A is invertible (equivalently, A has maximum rank n).

Example 5.26

Consider the set Seq $\mathbb{R}$ of all real sequences

$$(a_n)_{n \geqslant 1} \equiv a_1, a_2, a_3, \ldots, a_n, a_{n+1}, \ldots$$

of real numbers. We can make Seq $\mathbb{R}$ into a real vector space in an obvious way, namely by defining an addition and a multiplication by scalars as follows:

$$(a_n)_{n \geqslant 1} + (b_n)_{n \geqslant 1} = (a_n + b_n)_{n \geqslant 1};$$
$$\lambda (a_n)_{n \geqslant 1} = (\lambda a_n)_{n \geqslant 1}.$$

Define a sequence to be **finite** if there is some element a_m of the sequence such that $a_p = 0$ for all $p > m$; put another way, if there are only finitely many non-zero elements in the sequence.

The set $\mathrm{Seq}_f\ \mathbb{R}$ of all finite sequences is clearly closed under the above addition and multiplication by scalars and so is a subspace of $\mathrm{Seq}\ \mathbb{R}$. Consider the (finite) sequences that are represented as follows:

$$e_1 \equiv 1,0,0,0,\ldots$$

$$e_2 \equiv 0,1,0,0,\ldots$$

$$e_3 \equiv 0,0,1,0,\ldots$$

$$\vdots$$

$$e_i \equiv \underbrace{0,0,0,0,\ldots,0,}_{i-1}1,0,0,\ldots$$

$$\vdots$$

Clearly, every finite sequence can be expressed in a unique way as a linear combination of $e_1, e_2, e_3, \ldots$. Consequently, $\{e_1, e_2, e_3, \ldots\}$ forms a basis for the subspace $\mathrm{Seq}_f\ \mathbb{R}$ of finite sequences.

Note that this basis is *infinite*.

EXERCISES

5.20 Determine which of the following are bases of $\mathbb{R}^3$:
   (1) $\{(1,1,1),(1,2,3),(2,-1,1)\}$;
   (2) $\{(1,1,2),(1,2,5),(5,3,4)\}$.

5.21 Show that

$$\{(1,1,0,0),(-1,-1,1,2),(1,-1,1,3),(0,1,-1,-3)\}$$

   is a basis of $\mathbb{R}^4$ and express a general vector (a,b,c,d) as a linear combination of these basis elements.

5.22 Show that

$$X = \left\{ \begin{bmatrix} 1 \\ 2 \\ 1 \end{bmatrix}, \begin{bmatrix} 1 \\ 2 \\ 3 \end{bmatrix} \right\}, \qquad Y = \left\{ \begin{bmatrix} 0 \\ 0 \\ 1 \end{bmatrix}, \begin{bmatrix} 1 \\ 2 \\ 5 \end{bmatrix} \right\}$$

   are bases for the same subspace of $\mathrm{Mat}_{3\times 1}\ \mathbb{R}$.

Our objective now is to prove that if a vector space V has a finite basis B then every basis of V is finite and has the same number of elements as B. This is a consequence of the following result.

Theorem 5.7

Let V be a vector space that is spanned by the finite set $G = \{v_1,\ldots,v_n\}$. If $I = \{w_1,\ldots,w_m\}$ is a linearly independent subset of V then necessarily $m \leqslant n$.

Proof

Consider $w_1 \in I$. Since G is a spanning set of V, there exist scalars $\lambda_1, \ldots, \lambda_n$ such that

$$w_1 = \lambda_1 v_1 + \cdots + \lambda_n v_n$$

and at least one of the λ_i is non-zero (otherwise every $\lambda_i = 0$ whence $w_1 = 0_V$ and this contradicts Theorem 5.4). By a suitable change of indices if necessary, we may assume without loss that $\lambda_1 \neq 0$. We then have

$$v_1 = \lambda_1^{-1} w_1 - \lambda_1^{-1} \lambda_2 v_2 - \ldots - \lambda_1^{-1} \lambda_n v_n,$$

which shows that

$$V = \text{Span } G = \text{Span}\{v_1, v_2, \ldots, v_n\} \subseteq \text{Span}\{w_1, v_2, v_3, \ldots, v_n\}.$$

It follows that

$$V = \text{Span}\{w_1, v_2, v_3, \ldots, v_n\}.$$

Now w_2 can be written as a linear combination of $w_1, v_2, \ldots, v_n$ in which at least one of the coefficients of the v_j is non-zero (otherwise w_2 is a linear combination of w_1, a contradiction). Repeating the above argument we therefore obtain

$$V = \text{Span}\{w_1, w_2, v_3, \ldots, v_n\}.$$

Continuing in this way, we see that if $p = \min\{m, n\}$ then

$$V = \text{Span}\{w_1, \ldots, w_p, v_{p+1}, \ldots, v_n\}.$$

Now we see that $m > n$ is impossible; for in this case $p = n$ and we would have $V = \text{Span}\{w_1, \ldots, w_n\}$ whence the elements $w_{n+1}, \ldots, w_m$ would be linear combinations of $w_1, \ldots, w_n$ and this would contradict the fact that I is independent. Thus we conclude that $m \leqslant n$. $\square$

Corollary

If V has a finite basis B then every basis of V is finite and has the same number of elements as B.

Proof

Suppose that $B^\star$ were an infinite basis of V. Since, clearly, every subset of a linearly independent set is also linearly independent, every subset of $B^\star$ is linearly independent. Now $B^\star$, being infinite, contains finite subsets that have more elements than B. There would therefore exist a finite independent subset having more elements than B. Since this contradicts Theorem 5.7, we conclude that all bases of V must be finite.

Suppose now that the basis B has n elements and let $B^\star$ be a basis with $n^\star$ elements. By Theorem 5.7, we have $n^\star \leqslant n$. But, inverting the rôles of B and $B^\star$, we deduce also that $n \leqslant n^\star$. Thus $n^\star = n$ and so all bases have the same number of elements. $\square$

Because of the above result, we can introduce the following notion.

Definition

By a **finite-dimensional** vector space we shall mean a vector space V that has a finite basis. The number of elements in any basis of V is called the **dimension** of V and will be denoted by dim V.

Example 5.27

The vector space $\mathbb{R}^n$ has dimension n. In fact, as we have seen before, $\{e_1, \ldots, e_n\}$ is a basis.

Example 5.28

The vector space $\text{Mat}_{m \times n} \, \mathbb{R}$ is of dimension mn. To see this, observe that if E_{ij} is the $m \times n$ matrix that has 1 in the (i, j)-th position and 0 elsewhere then

$$\{E_{ij} \; ; \; i = 1, \ldots, m, \; j = 1, \ldots, n\}$$

is a basis for $\text{Mat}_{m \times n} \, \mathbb{R}$.

Example 5.29

The vector space $\mathbb{R}_n[X]$ of real polynomials of degree at most n is of dimension $n + 1$. In fact,

$$\{1, X, X^2, \ldots, X^n\}$$

is a basis for this space.

Example 5.30

The set V of complex matrices of the form

$$\begin{bmatrix} \alpha & \beta \\ \gamma & -\alpha \end{bmatrix}$$

forms a real vector space of dimension 6.

In fact, V is a subspace of the real vector space $\text{Mat}_{2 \times 2} \, \mathbb{C}$. Morover, the matrix

$$\begin{bmatrix} \alpha & \beta \\ \gamma & -\alpha \end{bmatrix} = \begin{bmatrix} a + ib & c + id \\ e + if & -a - ib \end{bmatrix}$$

can be written as

$$a \begin{bmatrix} 1 & 0 \\ 0 & -1 \end{bmatrix} + b \begin{bmatrix} i & 0 \\ 0 & -i \end{bmatrix} + c \begin{bmatrix} 0 & 1 \\ 0 & 0 \end{bmatrix} + d \begin{bmatrix} 0 & i \\ 0 & 0 \end{bmatrix} + e \begin{bmatrix} 0 & 0 \\ 1 & 0 \end{bmatrix} + f \begin{bmatrix} 0 & 0 \\ i & 0 \end{bmatrix}$$

and as the six matrices involved in this linear combination belong to V and are clearly linearly independent over $\mathbb{R}$, they form a basis of the subspace that they span, which is V.

Example 5.31

The set W of complex matrices of the form

$$\begin{bmatrix} \alpha & \beta \\ -\bar{\beta} & -\bar{\alpha} \end{bmatrix}$$

is a real vector space of dimension 4.

In fact, W is a subspace of the real vector space $\text{Mat}_{2\times2}\,\mathbb{C}$. Morover, the matrix

$$\begin{bmatrix} \alpha & \beta \\ -\bar{\beta} & -\bar{\alpha} \end{bmatrix} = \begin{bmatrix} a+ib & c+id \\ -c+id & -a-ib \end{bmatrix}$$

can be written as

$$a\begin{bmatrix} 1 & 0 \\ 0 & -1 \end{bmatrix} + b\begin{bmatrix} i & 0 \\ 0 & -i \end{bmatrix} + c\begin{bmatrix} 0 & 1 \\ -1 & 0 \end{bmatrix} + d\begin{bmatrix} 0 & i \\ i & 0 \end{bmatrix}$$

and as the four matrices involved in this linear combination belong to W and are clearly linearly independent over $\mathbb{R}$, they form a basis of the subspace that they span, which is W.

EXERCISES

5.23 Let V be a vector space of dimension 2. If $\{v_1, v_2\}$ is a basis of V and if $w_1, w_2 \in V$ prove that the following statements are equivalent:

(1) $\text{Span}\{w_1, w_2\} = V$;

(2) there is an invertible matrix A such that

$$\begin{bmatrix} w_1 \\ w_2 \end{bmatrix} = A\begin{bmatrix} v_1 \\ v_2 \end{bmatrix}.$$

5.24 In the vector space $\mathbb{R}^4$ let

$$A = \text{Span}\{(1, 2, 0, 1), (-1, 1, 1, 1)\},$$
$$B = \text{Span}\{(0, 0, 1, 1)\}, (2, 2, 2, 2)\}.$$

Determine $A \cap B$ and compute its dimension.

5.25 If V is a vector space over $\mathbb{C}$ of dimension n, prove that V can be regarded as a vector space over $\mathbb{R}$ of dimension $2n$.

[*Hint.* Consider $\{v_1, \ldots, v_n, iv_1, \ldots, iv_n\}$ where $\{v_1, \ldots, v_n\}$ is a basis over $\mathbb{C}$.]

The reader will recall that the notion of linear independence was defined for a *non-empty* subset of a vector space. Now it is convenient to extend to the empty set $\emptyset$ the courtesy of being linearly independent, the justification for this being that the condition for a set of elements to be linearly independent can be viewed as being satisfied 'vacuously' by $\emptyset$.

Since we know that the smallest subspace of V is the singleton $\{0_V\}$, and since this is clearly the smallest subspace to contain $\emptyset$, we can also regard the zero subspace as being spanned by $\emptyset$.

These courtesies concerning $\emptyset$ mean that we can usefully regard $\emptyset$ as a basis for the zero subspace, which we can then say has dimension 0. This we shall do henceforth.

We shall now establish some important facts concerning bases.

Theorem 5.8

Let V be a finite-dimensional vector space. If G is a finite spanning set of V and if I is a linearly independent subset of V such that $I \subseteq G$ then there is a basis B of V such that $I \subseteq B \subseteq G$.

Proof

Observe first that if I also spans V then I is a basis of V and there is nothing to prove.

Suppose then that $V \neq \text{Span } I$. Then we must have $I \subset G$ (for otherwise $I = G$ and is a spanning set of V).

We note first that there exists $g_1 \in G \backslash I$ such that $g_1 \notin \text{Span } I$; for otherwise every element of $G \backslash I$ belongs to Span I whence $V = \text{Span } G \subseteq \text{Span } I$ and we have the contradiction $V = \text{Span } I$. We then observe that $I \cup \{g_1\}$ is linearly independent; otherwise we have the contradiction $g_1 \in \text{Span } I$.

Now if $I \cup \{g_1\}$ spans V then it is a basis, in which case no more proof is required since we can take $B = I \cup \{g_1\}$. If $I \cup \{g_1\}$ does not span V then we can repeat the above argument to produce an element $g_2 \in G \backslash (I \cup \{g_1\})$ with $I \cup \{g_1, g_2\}$ linearly independent.

Proceeding in this way we see, since G is finite by hypothesis, that for some m the set $B = I \cup \{g_1, g_2, \ldots, g_m\}$ is a basis of V with $I \subset B \subseteq G$. $\square$

Corollary 1

Every linearly independent subset I of a finite-dimensional vector space V can be extended to form a basis.

Proof

Take $G = I \cup B$ where B is any basis of V. Then by the above there is a basis $B^\star$ with $I \subseteq B^\star \subseteq I \cup B$. $\square$

Corollary 2

If V is of dimension n then every linearly independent set consisting of n elements is a basis of V.

Proof

Immediate by Corollary 1 and the Corollary to Theorem 5.7. $\square$

Corollary 3

If S is a subset of V then the following statements are equivalent:

(1) *S is a basis;*

(2) *S is a maximal independent subset* (in the sense that if I is an independent subset with $S \subseteq I$ then $S = I$);

(3) *S is a minimal spanning set* (in the sense that if G spans V and $G \subseteq S$ then $G = S$).

Proof

$(1) \Rightarrow (2)$: If I is independent with $S \subseteq I$ then by Corollary 1 there is a basis B such that $I \subseteq B$. Since S is a basis, and since all bases have the same number of elements, we deduce that $S = B = I$.

$(2) \Rightarrow (1)$: By Corollary 1 there is a basis B with $S \subseteq B$. But B is also independent so, by (2), we have $S = B$ and therefore S is a basis.

$(1) \Rightarrow (3)$: If G spans V then (recalling that $\emptyset$ is independent) there is a basis B with $\emptyset \subseteq B \subseteq G$. If $G \subseteq S$ then $B \subseteq S$ and both are bases. Again since bases have the same number of elements, we deduce that $B = G = S$.

$(3) \Rightarrow (1)$: There is a basis B with $\emptyset \subseteq B \subseteq S$. But B also spans V so, by (3), we have $B = S$ and so S is a basis. $\square$

Corollary 4

If V is of dimension n then every subset containing more than n elements is linearly dependent. No subset containing fewer than n elements can span V.

Proof

This is immediate from Corollary 3. $\square$

As for subspaces of finite-dimensional vector spaces, we have the following consequence.

Theorem 5.9

Let V be a finite-dimensional vector space. If W is a subspace of V then W is also of finite dimension, and

$$\dim W \leqslant \dim V.$$

Moreover, we have

$$\dim W = \dim V \iff W = V.$$

Proof

Suppose that V is of dimension n. If I is a linearly independent subset of W then, by Theorem 5.7, I has at most n elements. A maximal such subset B is then, by Corollary 3 of Theorem 5.8, a basis of W. Hence W is also of finite dimension, and $\dim W \leqslant \dim V$.

Finally, if dim $W = $ dim $V = n$ then B is a linearly independent subset of V having n elements whence, by Corollary 2 of Theorem 5.8, B is a basis of V. Hence $W = \text{Span } B = V$.   □

Example 5.32

Consider the real vector space $\mathbb{R}^2$. This is of dimension 2 and so if W is a subspace of $\mathbb{R}^2$ then by Theorem 5.9 the dimension of W is either 0, 1, or 2.

If dim $W = 0$ then we have $W = \{(0,0)\}$.

If dim $W = 2$ then, by Theorem 5.9, we have $W = \mathbb{R}^2$.

If dim $W = 1$ then W has a basis of a single non-zero element (x, y), so that

$$W = \{\lambda(x, y) \; ; \; \lambda \in \mathbb{R}\} = \{(\lambda x, \lambda y) \; ; \; \lambda \in \mathbb{R}\},$$

which is none other than the line passing through the origin $(0,0)$ and the point (x, y).

Example 5.33

Arguing in a similar way to the above, we can show that the subspaces of $\mathbb{R}^3$, corresponding to the dimensions $0, 1, 2, 3$, are:

the zero subspace $\{(0,0,0)\}$;

any line through the origin;

any plane through the origin;

$\mathbb{R}^3$ itself.

Example 5.34

If V is a vector space with dim $V = 10$ and X, Y are subspaces of V with dim $X = 8$ and dim $Y = 9$ then the smallest possible value of dim $(X \cap Y)$ is 7.

To see this, begin with a basis $\{v_1, \ldots, v_p\}$ of $X \cap Y$ and, using Corollary 1 of Theorem 5.8, extend this on the one hand to a basis

$$\{v_1, \ldots, v_p, v_{p+1}, \ldots, v_8\}$$

of X, and on the other hand to a basis

$$\{v_1, \ldots, v_p, w_{p+1}, \ldots, w_9\}$$

of Y. Observe that none of $w_{p+1}, \ldots, w_9$ belongs to X (for otherwise it belongs to $X \cap Y$, a contradiction), and so

$$\{v_1, \ldots, v_p, v_{p+1}, \ldots, v_8, w_{p+1}, \ldots, w_9\}$$

is linearly independent (otherwise one of the vectors $w_{p+1}, \ldots, w_9$ would belong to Span $\{v_1, \ldots, v_8\} = X$, a contradiction). Since dim $V = 10$, this set contains at most 10 elements. For this we must have $p \geqslant 7$.

To see that this lower bound of 7 is attainable, consider $V = \mathbb{R}^{10}$ and take for X the subspace consisting of those 10-tuples whose first and third components are 0, and for Y the subspace consisting of those 10-tuples whose second component is 0.

EXERCISES

5.26 Consider the subset of $\mathbb{R}^4$ given by
$$X = \{(2,2,1,3), (7,5,5,5), (3,2,2,1), (2,1,2,1)\}.$$
Find a basis of Span X and extend this to a basis of $\mathbb{R}^4$.

5.27 Find a basis for $\text{Mat}_{3\times 1} \mathbb{R}$ that contains both the vectors
$$\begin{bmatrix} 1 \\ 0 \\ 2 \end{bmatrix}, \quad \begin{bmatrix} 0 \\ 2 \\ 4 \end{bmatrix}.$$

5.28 Find a basis of $\mathbb{R}_3[X]$ containing the polynomials $1 + X + X^2$ and $X - X^3$.

SUPPLEMENTARY EXERCISES

5.29 For each of the following statements give a proof if it is true and a counter-example if it is false:

(1) If V is a vector space over a field F then a non-empty subset W of V is a subspace of V if and only if
$$(\forall x, y \in V)(\forall \lambda, \mu \in F) \qquad \lambda x + \mu y \in W.$$

(2) The subspace $\{(x, x, x) \; ; \; x \in \mathbb{R}\}$ of $\mathbb{R}^3$ is of dimension 3.

(3) Every spanning set contains a basis.

(4) The subspace of $\mathbb{R}^3$ spanned by $\{(1,2,1), (2,2,1)\}$ is

(a) $\{(a + 2b, 2a + 2b, a + b) \; ; \; a, b \in \mathbb{R}\}$.

(b) $\{(x + y, 2y, y) \; ; \; x, y \in \mathbb{R}\}$;

(5) If P, Q are subspaces of a finite dimensional vector space then

(a) $P \subseteq Q$ implies $\dim P \leqslant \dim Q$;

(b) $\dim P \leqslant \dim Q$ implies $P \subseteq Q$.

(6) If $\{x, y, z\}$ is a basis of $\mathbb{R}^3$ and w is a non-zero vector in $\mathbb{R}^3$ then $\{w + x, y, z\}$ is also a basis of $\mathbb{R}^3$.

5.30 Determine whether or not the following subsets of $\mathbb{R}^4$ are subspaces:

(1) $\{(a, b, c, d) \; ; \; a + b = c + d\}$;

(2) $\{(a, b, c, d) \; ; \; a + b = 1\}$;

(3) $\{(a, b, c, d) \; ; \; a^2 + b^2 = 0\}$;

(4) $\{(a, b, c, d) \; ; \; a^2 + b^2 = 1\}$.

5.31 Determine whether or not the following subsets of $\mathbb{R}^4$ are subspaces:

(1) $\{(x + 2y, 0, 2x - y, y) \; ; \; x, y \in \mathbb{R}\}$;

(2) $\{(x + 2y, x, 2x - y, y) \; ; \; x, y \in \mathbb{R}\}$.

5.32 Prove that the subset

$$\{(3 - i,\, 2 + 2i,\, 4),\, (2,\, 2 + 4i,\, 3),\, (1 - i,\, -2i,\, -1)\}$$

is a basis of the complex vector space $\mathbb{C}^3$.

Express each of the vectors $(1, 0, 0)$, $(0, 1, 0)$, $(0, 0, 1)$ as a linear combination of these basis vectors.

5.33 Find a basis for the solution space of the homogeneous system

$$\begin{bmatrix} 1 & 2 & 2 & 1 & -1 \\ 0 & 2 & 2 & -1 & -2 \\ 2 & 6 & 2 & 1 & -4 \\ 1 & 4 & 0 & 0 & -3 \end{bmatrix} \begin{bmatrix} x \\ y \\ z \\ t \\ w \end{bmatrix} = \begin{bmatrix} 0 \\ 0 \\ 0 \\ 0 \end{bmatrix}.$$

5.34 Let V be a finite-dimensional vector space. If A, B are subspaces of V, prove that so also is the set

$$A + B = \{a + b \;;\; a \in A,\, b \in B\}.$$

Prove further that if C is any subspace of V such that $A \subseteq C$ and $B \subseteq C$ then $A + B \subseteq C$ (in other words, $A + B$ is the smallest subspace of V that contains both A and B).

If L, M, N are subspaces of V prove that

$$L \cap [M + (L \cap N)] = (L \cap M) + (L \cap N).$$

Give an example to show that in general

$$L \cap (M + N) \neq (L \cap M) + (L \cap N).$$

5.35 Let n be a positive integer and let E_n be the set of mappings $f : \mathbb{R} \to \mathbb{R}$ that are given by a prescription of the form

$$f(x) = a_0 + \sum_{k=1}^{n} (a_k \cos kx + b_k \sin kx)$$

where $a_k, b_k \in \mathbb{R}$ for each k.

Prove that E_n is a subspace of $\mathrm{Map}\,(\mathbb{R}, \mathbb{R})$.

If $f \in E_n$ is the zero mapping, prove that all the coefficients a_k, b_k must be 0.

[*Hint.* Proceed by induction. For this, find a prescription for $\mathrm{D}^2 f + n^2 f$.]

Deduce that the $2n + 1$ functions

$$x \mapsto 1, \quad x \mapsto \cos kx, \quad x \mapsto \sin kx \quad (k = 1, \ldots, n)$$

form a basis for E_n.

5.36 Let $\alpha, \beta \in \mathbb{R}$ with $\alpha \neq \beta$ and let r, s be fixed positive integers. Show that the set of rational functions $f : \mathbb{R} \to \mathbb{R}$ given by a prescription of the form

$$f(x) = \frac{a_0 + a_1 x + \cdots + a_{r+s-1} x^{r+s-1}}{(x - \alpha)^r (x - \beta)^s}$$

where each $a_i \in \mathbb{R}$, is a subspace of $\text{Map}(\mathbb{R}, \mathbb{R})$ of dimension $r + s$.

[*Hint.* Show that the functions

$$x \mapsto f_i(x) = \frac{x^i}{(x - \alpha)^r (x - \beta)^s}$$

for $i = 0, \ldots, r + s - 1$ constitute a basis.]

Show also that if g_i and h_j are given by

$$g_i(x) = (x - \alpha)^i, \qquad h_j(x) = (x - \beta)^j$$

then

$$B = \{g_1, \ldots, g_r\} \cup \{h_1, \ldots, h_s\}$$

is also a basis.

[*Hint.* It suffices to prove that B is independent.]

5.37 For each positive integer k let $f_k : \mathbb{R} \to \mathbb{R}$ be given by

$$f_k(x) = \exp r_k x$$

where each $r_k \in \mathbb{R}$. Prove that $\{f_1, \ldots, f_n\}$ is linearly independent if and only if $r_1, \ldots, r_n$ are distinct.

5.38 Let $P_0(X), P_1(X), \ldots, P_n(X)$ be polynomials in $\mathbb{R}_n[X]$ such that, for each i, the degree of $P_i(X)$ is i. Prove that

$$\{P_0(X), P_1(X), \ldots, P_n(X)\}$$

is a basis of $\mathbb{R}_n[X]$.

5.39 A **net** over the closed interval $[0, 1]$ of $\mathbb{R}$ is a finite sequence $(a_i)_{0 \leqslant i \leqslant n+1}$ such that

$$0 = a_0 < a_1 < \cdots < a_n < a_{n+1} = 1.$$

A **step function** on the semi-open interval $[0, 1[$ is a map $f : [0, 1[\to \mathbb{R}$ for which there is a net $(a_i)_{0 \leqslant i \leqslant n+1}$ over $[0, 1]$ and a finite sequence $(b_i)_{0 \leqslant i \leqslant n}$ such that

$$(\forall x \in [a_i, a_{i+1}[) \qquad f(x) = b_i.$$

Sketch a picture of a step function.

Show that the set E of step functions on $[0, 1[$ is a vector space.

Show also that a basis of E is the set of functions $e_k : [0, 1[\to \mathbb{R}$ given by

$$e_k(x) = \begin{cases} 0 & \text{if } 0 \leqslant x < k; \\ 1 & \text{if } k \leqslant x < 1. \end{cases}$$

A **piecewise linear function** on $[0, 1[$ is a map $f : [0, 1[\to \mathbb{R}$ for which there is a net $(a_i)_{0 \leqslant i \leqslant n+1}$ and finite sequences $(b_i)_{0 \leqslant i \leqslant n}$, $(c_i)_{0 \leqslant i \leqslant n}$ such that

$$(\forall x \in [a_i, a_{i+1}[) \qquad f(x) = b_i x + c_i.$$

Sketch a picture of a piecewise linear function.

Show that the set F of piecewise linear functions on $[0, 1[$ is a vector space.

If G is the subset of F consisting of those piecewise linear functions g that are continuous with $g(0) = 0$, show that G is a subspace of F.

Show also that a basis of G is the set of functions $g_k : [0, 1[\to \mathbb{R}$ given by

$$g_k(x) = \begin{cases} 0 & \text{if } 0 \leqslant x < k; \\ x - k & \text{if } k \leqslant x < 1. \end{cases}$$

Show finally that every $f \in F$ can be written uniquely in the form $g + e$ where $g \in G$ and $e \in E$.

6.
Linear Mappings

In the study of any algebraic structure there are two concepts that are of paramount importance. The first is that of a **substructure** (i.e. a subset with the same type of structure), and the second is that of a **morphism** (i.e. a mapping from one structure to another of the same kind that is 'structure-preserving').

So far, we have encountered the notion of a substructure for a vector space; this is called a *subspace*. In this chapter we shall consider the notion of a morphism between vector spaces, i.e. a mapping from one vector space to another that is 'structure-preserving' in the following sense.

Definition

If V and W are vector spaces over the same field F then by a **linear mapping** (or **linear transformation**) from V to W we shall mean a mapping $f : V \to W$ such that

(1) $(\forall x, y \in V) \qquad f(x + y) = f(x) + f(y)$;

(2) $(\forall x \in V)(\forall \lambda \in F) \qquad f(\lambda x) = \lambda f(x)$.

- If $f : V \to W$ is linear then V is sometimes called the **departure space** and W the **arrival space** of f.

Example 6.1

The mapping $f : \mathbb{R}^2 \to \mathbb{R}^3$ given by

$$f(a, b) = (a + b, a - b, b)$$

is linear. In fact, for all (a, b) and (a', b') in $\mathbb{R}^2$ we have

$$
\begin{aligned}
f\big((a, b) + (a', b')\big) &= f(a + a', b + b') \\
&= (a + a' + b + b', a + a' - b - b', b + b') \\
&= (a + b, a - b, b) + (a' + b', a' - b', b') \\
&= f(a, b) + f(a', b')
\end{aligned}
$$

and, for all $(a, b) \in \mathbb{R}^2$ and all $\lambda \in \mathbb{R}$,

$$\begin{aligned} f\big(\lambda(a,b)\big) &= f(\lambda a, \lambda b) \\ &= (\lambda a + \lambda b, \lambda a - \lambda b, \lambda b) \\ &= \lambda(a + b, a - b, b) \\ &= \lambda f(a, b). \end{aligned}$$

Example 6.2

The mapping $\mathrm{pr}_i : \mathbb{R}^n \to \mathbb{R}$ described by

$$\mathrm{pr}_i(x_1, \dots, x_n) = x_i$$

(i.e. the mapping that picks out the i-th coordinate) is called the i-**th projection** of $\mathbb{R}^n$ onto $\mathbb{R}$. It is readily seen that (1) and (2) above are satisfied, so that pr_i is linear.

Example 6.3

Consider the differentiation map $D : \mathbb{R}_n[X] \to \mathbb{R}_n[X]$ given by

$$D(a_0 + a_1 X + \cdots + a_n X^n) = a_1 + 2a_2 X + \cdots + na_n X^{n-1}.$$

This mapping is linear; for if $p(X)$ and $q(X)$ are polynomials then we know from analysis that $D\big(p(X)+q(X)\big) = Dp(X)+Dq(X)$ and that, for every scalar λ, $D\big(\lambda p(X)\big) = \lambda\, Dp(X)$.

EXERCISES

6.1 Decide which of the following mappings $f : \mathbb{R}^3 \to \mathbb{R}^3$ are linear:
   (1) $f(x, y, z) = (y, z, 0)$;
   (2) $f(x, y, z) = (z, -y, x)$;
   (3) $f(x, y, z) = (|x|, -z, 0)$;
   (4) $f(x, y, z) = (x - 1, x, y)$;
   (5) $f(x, y, z) = (x + y, z, 0)$;
   (6) $f(x, y, z) = (2x, y - 2, 4y)$.

6.2 Let $B \in \mathrm{Mat}_{n \times n} \mathbb{R}$ be fixed and non-zero. Which of the following mappings $T_B : \mathrm{Mat}_{n \times n} \mathbb{R} \to \mathrm{Mat}_{n \times n} \mathbb{R}$ are linear?
   (1) $T_B(X) = XB - BX$;
   (2) $T_B(X) = XB^2 + BX$;
   (3) $T_B(X) = XB^2 - BX^2$.

6.3 Which of the following mappings are linear?
   (1) $f : \mathbb{R}_n[X] \to \mathbb{R}_3[X]$ given by $f\big(p(X)\big) = p(0)X^2 + Dp(0)X^3$;
   (2) $f : \mathbb{R}_n[X] \to \mathbb{R}_{n+1}[X]$ given by $f\big(p(X)\big) = p(0) + Xp(X)$;
   (3) $f : \mathbb{R}_n[X] \to \mathbb{R}_{n+1}[X]$ given by $f\big(p(X)\big) = 1 + Xp(X)$.

6.4 Let A be a given real $m \times n$ matrix. Prove that the mapping

$$f_A : \text{Mat}_{n \times 1} \mathbb{R} \to \text{Mat}_{m \times 1} \mathbb{R}$$

described by $f_A(\mathbf{x}) = A\mathbf{x}$ is linear.

6.5 Let $I : \mathbb{R}_n[X] \to \mathbb{R}$ be the **integration map** defined by

$$I(p(X)) = \int_0^1 p(X).$$

Prove that I is linear.

6.6 For an $m \times n$ matrix A let $f(A)$ be the Hermite form of A if $A \neq 0$, and let $f(0) = 0$. Is f linear?

The following result contains two important properties of linear mappings that will be used constantly in what follows.

Theorem 6.1

If the mapping $f : V \to W$ is linear then
   (1) $f(0_V) = 0_W$;
   (2) $(\forall x \in V) \quad f(-x) = -f(x)$.

Proof

   (1) We have $f(0_V) = f(0_F 0_V) = 0_F f(0_V) = 0_W$.
   (2) Using (1) we have, for every $x \in V$,

$$f(x) + f(-x) = f[x + (-x)] = f(0_V) = 0_W$$

from which the result follows on adding $-f(x)$ to each side. $\square$

EXERCISES

6.7 Let $B \in \text{Mat}_{n \times n} \mathbb{R}$ be fixed and non-zero. Prove that the mapping $T_B :$ $\text{Mat}_{n \times n} \mathbb{R} \to \text{Mat}_{n \times n} \mathbb{R}$ given by

$$T_B(A) = (A + B)^2 - (A + 2B)(A - 3B)$$

is linear if and only if $B^2 = 0$.

We shall now consider some important subsets that are associated with linear mappings. For this purpose we introduce the following notation.

If $f : V \to W$ is linear then for every subset X of V we define $f^{\to}(X)$ to be the subset of W given by

$$f^{\to}(X) = \{f(x) ; \ x \in X\};$$

and for every subset Y of W we define $f^{\leftarrow}(Y)$ to be the subset of V given by

$$f^{\leftarrow}(Y) = \{x \in V ; \ f(x) \in Y\}.$$

We often call $f^{\to}(X)$ the **direct image** of X under f, and $f^{\leftarrow}(Y)$ the **inverse image** of Y under f.

- The reader should be warned that this is not 'standard' notation, in the sense that most authors write $f(X)$ for $f^{\rightarrow}(X)$, and $f^{-1}(Y)$ for $f^{\leftarrow}(Y)$. We introduce this notation in order to reserve the notation f^{-1} as the standard notation for the inverse of a bijection.

One advantage that this non-standard notation has to offer is that it gives a visually appealing reminder that $f^{\rightarrow}$ sends subsets of V to subsets of W, and $f^{\leftarrow}$ lifts back subsets of W to subsets of V.

EXERCISES

6.8 Consider the differentiation map $D : \mathbb{R}_n[X] \rightarrow \mathbb{R}_n[X]$. Describe the sets $D^{\rightarrow}(\mathbb{R}_n[X])$ and $D^{\leftarrow}(0)$.

6.9 Prove that $f^{\rightarrow} = f^{\rightarrow} \circ f^{\leftarrow} \circ f^{\rightarrow}$ and that $f^{\leftarrow} = f^{\leftarrow} \circ f^{\rightarrow} \circ f^{\leftarrow}$.

The mappings $f^{\rightarrow}$ and $f^{\leftarrow}$ are each **inclusion-preserving** in the sense that

(a) $X_1 \subseteq X_2 \;\Rightarrow\; f^{\rightarrow}(X_1) \subseteq f^{\rightarrow}(X_2)$.
For, if $y \in f^{\rightarrow}(X_1)$ then $y = f(x_1)$ where $x_1 \in X_1 \subseteq X_2$.

(b) $Y_1 \subseteq Y_2 \;\Rightarrow\; f^{\leftarrow}(Y_1) \subseteq f^{\leftarrow}(Y_2)$.
For, if $x \in f^{\leftarrow}(Y_1)$ then $f(x) \in Y_1 \subseteq Y_2$.

Moreover, each of these mappings carries sub*spaces* to sub*spaces*:

Theorem 6.2

Let $f : V \rightarrow W$ be linear. If X is a subspace of V then $f^{\rightarrow}(X)$ is a subspace of W; and if Y is a subspace of W then $f^{\leftarrow}(Y)$ is a subspace of V.

Proof

Observe first that if X is a subspace of V then we have $0_V \in X$ and therefore $0_W = f(0_V) \in f^{\rightarrow}(X)$. Thus $f^{\rightarrow}(X) \neq \emptyset$.
   If now $y_1, y_2 \in f^{\rightarrow}(X)$ then $y_1 = f(x_1)$ and $y_2 = f(x_2)$ for some $x_1, x_2 \in X$. Consequently, since X is a subspace of V,

$$y_1 + y_2 = f(x_1) + f(x_2) = f(x_1 + x_2) \in f^{\rightarrow}(X);$$

and, for every scalar λ,

$$\lambda y_1 = \lambda f(x_1) = f(\lambda x_1) \in f^{\rightarrow}(X).$$

Thus $f^{\rightarrow}(X)$ is a subspace of W.
   Suppose now that Y is a subspace of W. Observe that $f(0_V) = 0_W \in Y$ gives $0_V \in f^{\leftarrow}(Y)$, and therefore $f^{\leftarrow}(Y) \neq \emptyset$.
   If now $x_1, x_2 \in f^{\leftarrow}(Y)$ then $f(x_1), f(x_2) \in Y$ and therefore

$$f(x_1 + x_2) = f(x_1) + f(x_2) \in Y$$

whence $x_1 + x_2 \in f^{\leftarrow}(Y)$; and, for every scalar λ,

$$f(\lambda x_1) = \lambda f(x_1) \in Y$$

whence $\lambda x_1 \in f^{\leftarrow}(Y)$. Thus $f^{\leftarrow}(Y)$ is a subspace of V. $\square$

EXERCISES

6.10 Show that the subset X of polynomials in $\mathbb{R}_{2n}[X]$ all of whose odd co-efficients a_{2i+1} are zero forms a subspace of $\mathbb{R}_{2n}[X]$. Describe $D^{\rightarrow}(X)$ and $D^{\leftarrow}(X)$.

6.11 Show that the mapping $f : \mathbb{R}^2 \to \mathbb{R}^2$ given by

$$f(x, y) = (x + y, x - y)$$

is linear. For each subspace X of $\mathbb{R}^2$ describe $f^{\rightarrow}(X)$ and $f^{\leftarrow}(X)$.

6.12 Let $f : V \to W$ be linear. If X is a subspace of V and Y is a subspace of W, prove that

$$f^{\rightarrow}[X \cap f^{\leftarrow}(Y)] = f^{\rightarrow}(X) \cap Y.$$

Deduce that

$$f^{\rightarrow}(X) \subseteq Y \iff X \subseteq f^{\leftarrow}(Y).$$

Of particular importance relative to any linear mapping $f : V \to W$ are the biggest possible direct image and the smallest possible inverse image.

The former is $f^{\rightarrow}(V)$; it is called the **image** (or **range**) of f and is denoted by Im f.

The latter is $f^{\leftarrow}(\{0_W\})$; it is called the **kernel** (or **null-space**) of f and is denoted by Ker f.

Pictorially, these sets can be depicted as follows:

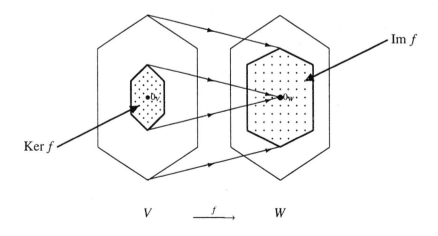

Example 6.4

Consider the i-th projection $\mathrm{pr}_i : \mathbb{R}^n \to \mathbb{R}$. Recall that $\mathrm{pr}_i(x_1, \ldots, x_n) = x_i$. The image of pr_i is therefore the whole of $\mathbb{R}$; and the kernel of pr_i is the set of n-tuples whose i-th component is 0.

Example 6.5

Consider the differentiation map $D : \mathbb{R}_n[X] \to \mathbb{R}_n[X]$. Its image is the set of all polynomials of degree at most $n - 1$; in other words, it is $\mathbb{R}_{n-1}[X]$. The kernel of D is the set of polynomials whose derivative is zero; in other words it is $\mathbb{R}$.

Example 6.6

If A is a given real $n \times n$ matrix, consider the linear mapping

$$f_A : \mathrm{Mat}_{n\times 1} \mathbb{R} \to \mathrm{Mat}_{n\times 1} \mathbb{R}$$

described by $f_A(\mathbf{x}) = A\mathbf{x}$. The image of f_A consists of all $n \times 1$ column matrices

$$\mathbf{y} = \begin{bmatrix} y_1 \\ \vdots \\ y_n \end{bmatrix} \text{ for which there exists } \mathbf{x} = \begin{bmatrix} x_1 \\ \vdots \\ x_n \end{bmatrix} \text{ such that } A\mathbf{x} = \mathbf{y}; \text{ i.e. the set of all } \mathbf{y} \text{ such}$$

that there exist $x_1, \ldots, x_n$ with

$$\mathbf{y} = x_1\mathbf{a}_1 + \cdots + x_n\mathbf{a}_n.$$

In other words, $\mathrm{Im}\, f_A$ is the subspace of $\mathrm{Mat}_{n\times 1} \mathbb{R}$ that is spanned by the columns of A.

As for the kernel of f_A, this is the subspace of $\mathrm{Mat}_{n\times 1} \mathbb{R}$ consisting of the column matrices $\mathbf{x}$ such that $A\mathbf{x} = \mathbf{0}$; i.e. the **solution space** of the system $A\mathbf{x} = \mathbf{0}$.

Example 6.7

Consider the subspace of $\mathrm{Map}(\mathbb{R}, \mathbb{R})$ that is given by

$$V = \mathrm{Span}\{\sin, \cos\},$$

i.e. the set of all real functions f given by a prescription of the form

$$f(x) = a \sin x + b \cos x.$$

Let $I : V \to \mathbb{R}$ be given by

$$I(f) = \int_0^\pi f.$$

Using basic properties of integrals, we see that I is linear.

Now if $f(x) = a \sin x + b \cos x$ then $f \in \mathrm{Ker}\, I$ if and only if

$$\int_0^\pi (a \sin x + b \cos x)dx = 0.$$

This is the case if and only if $a = 0$. Consequently, we see that $\mathrm{Ker}\, I = \mathrm{Span}\{\cos\}$.

Example 6.8

Consider the mapping $f : \mathbb{R}^4 \to \mathbb{R}^3$ given by

$$f(a,b,c,d) = (a+b, b-c, a+d).$$

Since

$$(a+b, b-c, a+d) = a(1,0,1) + b(1,1,0) + c(0,-1,0) + d(0,0,1)$$

we see that

$$\text{Im } f = \text{Span}\{(1,0,1),(1,1,0),(0,-1,0),(0,0,1)\}.$$

To find a basis for Im f, proceed as follows. Observe that Im f is the subspace spanned by the rows of the matrix

$$A = \begin{bmatrix} 1 & 0 & 1 \\ 1 & 1 & 0 \\ 0 & -1 & 0 \\ 0 & 0 & 1 \end{bmatrix}.$$

The Hermite form of A is

$$\begin{bmatrix} 1 & 0 & 0 \\ 0 & 1 & 0 \\ 0 & 0 & 1 \\ 0 & 0 & 0 \end{bmatrix}.$$

Since the rows of this matrix span the same subspace, and since they are linearly independent, we deduce that a basis for Im f is

$$\{(1,0,1),(1,1,0),(0,-1,0)\}.$$

Thus Im $f = \mathbb{R}^3$.

EXERCISES

6.13 Find Im f and Ker f when $f : \mathbb{R}^3 \to \mathbb{R}^3$ is given by

$$f(a,b,c) = (a+b, b+c, a+c).$$

6.14 If $f : \mathbb{R}^2 \to \mathbb{R}^2$ is given by $f(a,b) = (b,0)$, prove that Im $f = $ Ker f.

6.15 Give an example of a linear mapping for which Im $f \subset$ Ker f; and an example where Ker $f \subset$ Im f.

6.16 Let $f : \mathbb{R}^5 \to \mathbb{R}^4$ be given by

$$f(a,b,c,d,e) = (a-c+3d-e, a+2d-e, 2a-c+5d-e, -c+d).$$

Find bases for Im f and Ker f.

6.17 Let $f : \mathbb{R}_2[X] \to \mathbb{R}_3[X]$ be given by

$$f(p(X)) = X^2 \, Dp(X).$$

Prove that f is linear and determine bases for Im f and Ker f.

6.18 Consider the subspace A of $\mathbb{R}^4$ given by
$$A = \{(x, 0, z, 0) \; ; \; x, z \in \mathbb{R}\}.$$
Determine linear mappings $f, g : \mathbb{R}^4 \to \mathbb{R}^4$ such that
(1) Im $f = A$;
(2) Ker $g = A$.

Definition

A linear mapping $f : V \to W$ is said to be **surjective** if Im $f = W$ (in other words, if every element of W is the image under f of some element of V); and **injective** if $f(x) \neq f(y)$ whenever $x \neq y$ (in other words, if f carries distinct elements to distinct elements).

Example 6.9

The i-th projection $\text{pr}_i : \mathbb{R}^n \to \mathbb{R}$ is surjective but not injective.

Example 6.10

The linear mapping $f : \mathbb{R}^2 \to \mathbb{R}^3$ given by
$$f(x, y) = (y, 0, x)$$
is injective but not surjective.

Example 6.11

The differentiation map $D : \mathbb{R}_n[X] \to \mathbb{R}_n[X]$ is neither injective nor surjective.

From the above definition, a linear mapping $f : V \to W$ is surjective if Im f is as large as it can be, namely W. Dually, we can show that $f : V \to W$ is injective if Ker f is as small as it can be, namely $\{0_V\}$:

Theorem 6.3

If $f : V \to W$ is linear then the following statements are equivalent :
(1) *f is injective;*
(2) $\overline{\text{Ker } f = \{0\}}$.

Proof

$(1) \Rightarrow (2)$: Suppose that f is injective. Then f is such that
$$x \neq y \Rightarrow f(x) \neq f(y)$$
or, equivalently,
$$f(x) = f(y) \Rightarrow x = y.$$
Suppose now that $x \in$ Ker f. Then we have
$$f(x) = 0_W = f(0_V)$$

whence we see that $x = 0_V$ and consequently $\operatorname{Ker} f = \{0\}$.

$(2) \Rightarrow (1)$: Suppose that $\operatorname{Ker} f = \{0\}$ and let $f(x) = f(y)$. Then

$$f(x - y) = f[x + (-y)] = f(x) + f(-y) = f(x) - f(y) = 0_W$$

so that $x - y \in \operatorname{Ker} f = \{0_V\}$ and hence $x = y$, i.e. f is injective. $\square$

Example 6.12

The linear mapping $f : \mathbb{R}^3 \to \mathbb{R}^3$ given by

$$f(x, y, z) = (x + z, \ x + y + 2z, \ 2x + y + 3z)$$

is neither surjective nor injective.

In fact, we have that $(a, b, c) \in \operatorname{Im} f$ if and only if the system of equations

$$
\begin{array}{rcrcrcl}
x & & & + & z & = & a \\
x & + & y & + & 2z & = & b \\
2x & + & y & + & 3z & = & c
\end{array}
$$

is consistent. The augmented matrix of the system is

$$
\begin{bmatrix}
1 & 0 & 1 & a \\
1 & 1 & 2 & b \\
2 & 1 & 3 & c
\end{bmatrix}
$$

and this has Hermite form

$$
\begin{bmatrix}
1 & 0 & 1 & a \\
0 & 1 & 1 & b - a \\
0 & 0 & 0 & c - b - a
\end{bmatrix}.
$$

We deduce from this that $(a, b, c) \in \operatorname{Im} f$ if and only if $c = a + b$, whence f is not surjective.

Now $(x, y, z) \in \operatorname{Ker} f$ if and only if

$$
\begin{array}{rcrcrcl}
x & & & + & z & = & 0 \\
x & + & y & + & 2z & = & 0 \\
2x & + & y & + & 3z & = & 0
\end{array},
$$

which is the associated homogeneous system of equations. By Theorem 6.3, for $\operatorname{Ker} f$ to be the zero subspace we require this system to have a unique solution (namely the trivial solution $(0, 0, 0)$). But, from the above Hermite form, the coefficient matrix has rank 2 and so, by Theorem 3.16, non-trivial solutions exist. Hence f is not injective.

EXERCISES

6.19 Show that the linear mapping $f : \mathbb{R}^3 \to \mathbb{R}^3$ given by

$$f(x, y, z) = (x + y + z, \ 2x - y - z, \ x + 2y - z)$$

is both surjective and injective.

6.20 Prove that if the linear mapping $f : V \to W$ is injective and $\{v_1, \ldots, v_n\}$ is a linearly independent subset of V then $\{f(v_1), \ldots, f(v_n)\}$ is a linearly independent subset of W.

In the case of finite-dimensional vector spaces there is an important connection between the dimensions of the subspaces Im f and Ker f.

Theorem 6.4

[Dimension Theorem] *Let V and W be vector spaces of finite dimension over a field F. If $f : V \to W$ is linear then*

$$\dim V = \dim \text{Im } f + \dim \text{Ker } f.$$

Proof

Let $\{w_1, \ldots, w_m\}$ be a basis of Im f, and let $\{v_1, \ldots, v_n\}$ be a basis of Ker f. Since each $w_i \in$ Im f, we can choose $v_1^\star, \ldots, v_m^\star \in V$ such that $f(v_i^\star) = w_i$ for $i = 1, \ldots, m$. We shall show that

$$\{v_1^\star, \ldots, v_m^\star, v_1, \ldots, v_n\}$$

is a basis of V, whence the result follows.

Suppose that $x \in V$. Since $f(x) \in$ Im f there exist $\lambda_1, \ldots, \lambda_m \in F$ such that

$$f(x) = \sum_{i=1}^{m} \lambda_i w_i = \sum_{i=1}^{m} \lambda_i f(v_i^\star) = \sum_{i=1}^{m} f(\lambda_i v_i^\star) = f\left(\sum_{i=1}^{m} \lambda_i v_i^\star\right).$$

It follows that

$$x - \sum_{i=1}^{m} \lambda_i v_i^\star \in \text{Ker } f$$

and so there exist $\mu_1, \ldots, \mu_n \in F$ such that

$$x - \sum_{i=1}^{m} \lambda_i v_i^\star = \sum_{j=1}^{n} \mu_j v_j.$$

Thus every $x \in V$ is a linear combination of $v_1^\star, \ldots, v_m^\star, v_1, \ldots, v_n$ and so

$$V = \text{Span}\{v_1^\star, \ldots, v_m^\star, v_1, \ldots, v_n\}.$$

Suppose now that

(1)
$$\sum_{i=1}^{m} \lambda_i v_i^\star + \sum_{j=1}^{n} \mu_j v_j = 0.$$

Then we have

$$\sum_{i=1}^{m} \lambda_i v_i^\star = -\sum_{j=1}^{n} \mu_j v_j \in \text{Ker } f$$

and consequently

$$\sum_{i=1}^{m} \lambda_i w_i = \sum_{i=1}^{m} \lambda_i f(v_i^\star) = f\left(\sum_{i=1}^{m} \lambda_i v_i^\star\right) = 0$$

whence $\lambda_1 = \cdots = \lambda_m = 0$ since $\{w_1, \ldots, w_m\}$ is a basis of Im f. It now follows from (1) that $\sum_{j=1}^{n} \mu_j v_j = 0$ whence $\mu_1 = \cdots = \mu_n = 0$ since $\{v_1, \ldots, v_n\}$ is a basis of Ker f. Thus we see that the spanning set

$$\{v_1^\star, \ldots, v_m^\star, v_1, \ldots, v_n\}$$

is also linearly independent and is therefore a basis of V. $\square$

Definition

If f is a linear mapping then dim Im f is called the **rank** of f; and dim Ker f is called the **nullity** of f.

With this terminology, the dimension theorem above can be stated in the form:

$$rank + nullity = dimension\ of\ departure\ space.$$

Example 6.13

Consider $\mathrm{pr}_1 : \mathbb{R}^3 \to \mathbb{R}$ given by $\mathrm{pr}_1(x, y, z) = x$. We have Im $\mathrm{pr}_1 = \mathbb{R}$ which is of dimension 1 since $\{1\}$ is a basis of the real vector space $\mathbb{R}$; so pr_1 is of rank 1. Also, Ker pr_1 is the y, z-plane which is of dimension 2. Thus pr_1 is of nullity 2.

EXERCISES

6.21 Let V be a vector space of dimension $n \geqslant 1$. If $f : V \to V$ is linear, prove that the following statements are equivalent:

(1) Im f = Ker f;

(2) $f \neq 0, f^2 = 0$, n is even, and the rank of f is $\frac{1}{2}n$.

6.22 Give an example of a vector space V and a linear mapping $f : V \to V$ with the property that not every element of V can be written as the sum of an element of Im f and an element of Ker f.

6.23 In the vector space of real continuous functions let

$$W = \mathrm{Span}\{\sin, \cos\}.$$

Determine the nullity of $\vartheta : W \to \mathbb{R}$ when ϑ is given by

(1) $\vartheta(f) = \displaystyle\int_0^\pi f$;

(2) $\vartheta(f) = \displaystyle\int_0^{2\pi} f$;

(3) $\vartheta(f) = Df(0)$.

6.24 In the vector space of real continuous functions let

$$W = \text{Span}\,\{f, g, h\}$$

where

$$f(x) = e^x, \quad g(x) = e^{-x}, \quad h(x) = x.$$

Let $T : W \rightarrow W$ be the linear mapping given by

$$T(\vartheta) = D^2\vartheta - \vartheta.$$

Determine the rank and nullity of T.

As an application of the dimension theorem, we now establish another result that is somewhat surprising.

Theorem 6.5

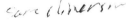

Let V and W be vector spaces each of dimension n over a field F. If $f : V \rightarrow W$ is linear then the following statements are equivalent:

(1) *f is injective*;
(2) *f is surjective*;
(3) *f is bijective*;
(4) *f carries bases to bases, in the sense that if $\{v_1, \ldots, v_n\}$ is a basis of V then* $\{f(v_1), \ldots, f(v_n)\}$ *is a basis of W.*

Proof

(1) $\Rightarrow$ (3) : Suppose that f is injective. Then $\text{Ker}\, f = \{0\}$ and so $\dim \text{Ker}\, f = 0$. By Theorem 6.4, it follows that

$$\dim \text{Im}\, f = n = \dim V = \dim W.$$

It now follows by Theorem 5.9 that $\text{Im}\, f = W$ and so f is also surjective, and hence is bijective.

(2) $\Rightarrow$ (3) : Suppose that f is surjective. Then $\text{Im}\, f = W$ and so, by Theorem 6.4,

$$\dim \text{Im}\, f = \dim W = n = \dim V = \dim \text{Im}\, f + \dim \text{Ker}\, f$$

whence $\dim \text{Ker}\, f = 0$. Thus $\text{Ker}\, f = \{0\}$ and so, by Theorem 6.3, f is also injective, and hence is bijective.

(3) $\Rightarrow$ (1) and (3) $\Rightarrow$ (2) are clear.

(1) $\Rightarrow$ (4) : Suppose that f is injective. If $\{v_1, \ldots, v_n\}$ is a basis of V then the elements $f(v_1), \ldots, f(v_n)$ are distinct. If now $\sum_{i=1}^{n} \lambda_i f(v_i) = 0$ then $f\left(\sum_{i=1}^{n} \lambda_i v_i\right) = 0$ and so, since $\text{Ker}\, f = \{0\}$, we have $\sum_{i=1}^{n} \lambda_i v_i = 0$ and hence $\lambda_1 = \cdots = \lambda_n = 0$. Thus $\{f(v_1), \ldots, f(v_n)\}$ is linearly independent. That it is now a basis follows from Corollary 2 of Theorem 5.8.

(4) $\Rightarrow$ (2) : Since every linear combination of $f(v_1), \ldots, f(v_n)$ belongs to Im f, it is clear from (4) that Im $f = W$ and so f is surjective. $\square$

Definition

A bijective linear mapping is called a **linear isomorphism**, or simply an **isomorphism**. We say that vector spaces V, W are **isomorphic**, and write $V \simeq W$, if there is an isomorphism $f : V \to W$.

Example 6.14

Let $A = \{(x, y, 0) \; ; \; x, y \in \mathbb{R}\}$ be the x, y-plane in $\mathbb{R}^3$, and let $B = \{(x, 0, z) \; ; \; x, z \in \mathbb{R}\}$ be the x, z-plane. Consider the mapping $f : A \to B$ given by

$$f(x, y, 0) = (x, 0, y).$$

Clearly, f is linear and bijective. Thus f is an isomorphism and so $A \simeq B$.

This example is a particular case of the following general situation.

Theorem 6.6

Let V be a vector space of dimension $n \geqslant 1$ over a field F. Then V is isomorphic to the vector space F^n.

Proof

Let $\{v_1, \ldots, v_n\}$ be a basis of V. Consider the mapping $f : V \to F^n$ given by the prescription

$$f\left(\sum_{i=1}^{n} \lambda_i v_i \right) = (\lambda_1, \ldots, \lambda_n).$$

Since for every $x \in V$ there are unique scalars $\lambda_1, \ldots, \lambda_n$ such that $x = \sum_{i=1}^{n} \lambda_i v_i$, it is clear that f is a bijection. It is clear that f is linear. Hence f is an isomorphism. $\square$

Corollary

If V and W are vector spaces of the same dimension n over F then V and W are isomorphic.

Proof

There are isomorphisms $f_V : V \to F^n$ and $f_W : W \to F^n$. Since the inverse of an isomorphism is clearly also an isomorphism, so then is the composite mapping $f_W^{-1} \circ f_V : V \to W$. $\square$

EXERCISES

6.25 Exhibit an isomorphism from $\mathbb{R}_n[X]$ to $\mathbb{R}^{n+1}$.

6.26 If $f : \mathbb{R} \to \mathbb{R}$ is linear and $\vartheta : \mathbb{R}^2 \to \mathbb{R}^2$ is defined by

$$\vartheta(x, y) = (x, y - f(x)),$$

prove that ϑ is an isomorphism.

Our next objective is to prove that a linear mapping enjoys the property of being completely and uniquely determined by its action on a basis. This is a consequence of the following result.

Theorem 6.7

Let V and W be vector spaces over a field F. If $\{v_1, \ldots, v_n\}$ is a basis of V and $w_1, \ldots, w_n$ are elements of W (not necessarily distinct) then there is a unique linear mapping $f : V \to W$ such that

$$(i = 1, \ldots, n) \qquad f(v_i) = w_i.$$

Proof

Since every element of V can be expressed uniquely in the form $\sum_{i=1}^{n} \lambda_i v_i$, we can define a mapping $f : V \to W$ by the prescription

$$f\left(\sum_{i=1}^{n} \lambda_i v_i\right) = \sum_{i=1}^{n} \lambda_i w_i,$$

i.e. taking x as a linear combination of the basis elements, define $f(x)$ to be the same linear combination of the elements $w_1, \ldots, w_n$.

It is readily verified that f is linear. Moreover, for each i, we have

$$f(v_i) = f\left(\sum_{j=1}^{n} \delta_{ij} v_j\right) = \sum_{j=1}^{n} \delta_{ij} w_j = w_i.$$

As for the uniqueness, suppose that $g : V \to W$ is also linear and such that $g(v_i) = w_i$ for each i. Given $x \in V$, say $x = \sum_{i=1}^{n} \lambda_i v_i$, we have

$$g(x) = g\left(\sum_{i=1}^{n} \lambda_i v_i\right) = \sum_{i=1}^{n} \lambda_i g(v_i) = \sum_{i=1}^{n} \lambda_i w_i = f(x)$$

whence $g = f$. $\square$

Corollary 1

A linear mapping is completely and uniquely determined by its action on a basis.

Proof

If $f : V \to W$ is linear and $B = \{v_1, \ldots, v_n\}$ is a basis of V let $w_i = f(v_i)$ for each i. Then by the above f is the only linear mapping that sends v_i to w_i. Moreover, knowing the action of f on the basis B, we can compute $f(x)$ for every x; for $x = \sum_{i=1}^{n} \lambda_i v_i$ gives $f(x) = \sum_{i=1}^{n} \lambda_i f(v_i)$. $\square$

Corollary 2

Two linear mappings $f, g : V \to W$ are equal if and only if they agree on any basis of V.

Proof

If $f(v_i) = g(v_i)$ for every basis element v_i then by the above uniqueness we have that $f = g$. $\square$

Example 6.15

Consider the basis $\{(1, 1, 0), (1, 0, 1), (0, 1, 1)\}$ of $\mathbb{R}^3$. If $f : \mathbb{R}^3 \to \mathbb{R}^2$ is linear and such that

$$f(1, 1, 0) = (1, 2), \quad f(1, 0, 1) = (0, 0), \quad f(0, 1, 1) = (2, 1),$$

then we can determine f completely.

In fact, we have

$$(1, 0, 0) = \tfrac{1}{2}(1, 1, 0) + \tfrac{1}{2}(1, 0, 1) - \tfrac{1}{2}(0, 1, 1)$$

and therefore

$$
\begin{aligned}
f(1, 0, 0) &= \tfrac{1}{2}f(1, 1, 0) + \tfrac{1}{2}f(1, 0, 1) - \tfrac{1}{2}f(0, 1, 1) \\
&= \tfrac{1}{2}(1, 2) + \tfrac{1}{2}(0, 0) - \tfrac{1}{2}(2, 1) \\
&= (-\tfrac{1}{2}, \tfrac{1}{2}).
\end{aligned}
$$

Likewise,

$$
\begin{aligned}
(0, 1, 0) &= \tfrac{1}{2}(1, 1, 0) - \tfrac{1}{2}(1, 0, 1) + \tfrac{1}{2}(0, 1, 1), \\
(0, 0, 1) &= -\tfrac{1}{2}(1, 1, 0) + \tfrac{1}{2}(1, 0, 1) + \tfrac{1}{2}(0, 1, 1)
\end{aligned}
$$

give

$$
\begin{aligned}
f(0, 1, 0) &= \tfrac{1}{2}(1, 2) + \tfrac{1}{2}(2, 1) = (\tfrac{3}{2}, \tfrac{3}{2}) \\
f(0, 0, 1) &= -\tfrac{1}{2}(1, 2) + \tfrac{1}{2}(2, 1) = (\tfrac{1}{2}, -\tfrac{1}{2}).
\end{aligned}
$$

Consequently, f is given by

$$
\begin{aligned}
f(x, y, z) &= f[x(1, 0, 0) + y(0, 1, 0) + z(0, 0, 1)] \\
&= xf(1, 0, 0) + yf(0, 1, 0) + zf(0, 0, 1) \\
&= x(-\tfrac{1}{2}, \tfrac{1}{2}) + y(\tfrac{3}{2}, \tfrac{3}{2}) + z(\tfrac{1}{2}, -\tfrac{1}{2}) \\
&= \left(\tfrac{1}{2}(-x + 3y + z), \tfrac{1}{2}(x + 3y - z)\right).
\end{aligned}
$$

Note that, alternatively, we could first have expressed (x, y, z) as a linear combination of the given basis elements by solving an appropriate system of equations, then using the given data.

Finally, let us note that Theorem 6.5 is not true for vector spaces of infinite dimension:

Example 6.16

Let $V = \text{Seq}_f \, \mathbb{R}$ be the infinite-dimensional vector space of finite sequences of real numbers described in Example 5.26. Since every element of V is a (finite) linear combination of basis elements, we can define a linear mapping $f : V \to V$ by specifying $f(e_i)$ for the basis elements $e_1, e_2, e_3, \ldots$ and extending to all of V by linearity.

Consider then the definition

$$f(e_i) = \begin{cases} 0 & \text{if } i \text{ is odd;} \\ e_{\frac{1}{2}i} & \text{if } i \text{ is even.} \end{cases}$$

Since $f(e_1) = 0 = f(e_3)$ we see that f is not injective. But, given any basis element e_n we have $e_n = f(e_{2n}) \in \text{Im } f$, so the subspace spanned by these elements (namely, the whole of V) is contained in $\text{Im } f$. Hence $\text{Im } f = V$ and so f is surjective.

If we define $g : V \to V$ by specifying $g(e_i) = e_{2i}$ for every i then we obtain an injective linear mapping that is not surjective.

EXERCISES

6.27 Show that $\{(1, 1, 1), (1, 2, 3), (1, 1, 2)\}$ is a basis of $\mathbb{R}^3$. If $f : \mathbb{R}^3 \to \mathbb{R}^3$ is linear and such that

$$f(1, 1, 1) = (1, 1, 1), \; f(1, 2, 3) = (-1, -2, -3), \; f(1, 1, 2) = (2, 2, 4),$$

determine $f(x, y, z)$ for all $(x, y, z) \in \mathbb{R}^3$.

6.28 If $f : \mathbb{R}_2[X] \to \mathbb{R}_3[X]$ is linear and such that $f(1) = 1$, $f(X) = X^2$ and $f(X^2) = X + X^3$, determine $f(a + bX + cX^2)$.

SUPPLEMENTARY EXERCISES

6.29 Let $f : \mathbb{C} \to \mathbb{C}$ be given by $f(x+iy) = x-iy$. Prove that if $\mathbb{C}$ is considered as a real vector space then f is linear, whereas if $\mathbb{C}$ is considered as a complex vector space f is not linear.

6.30 Let V be a vector space of dimension 3 with $\{v_1, v_2, v_3\}$ a basis. Let W be a vector space of dimension 2 with $\{w_1, w_2\}$ a basis. Let $f : V \to W$ be defined by

$$f(\lambda_1 v_1 + \lambda_2 v_2 + \lambda_3 v_3) = (\lambda_1 + \mu)w_1 + (\lambda_2 + \lambda_3)w_2.$$

Determine the values of μ for which f is linear. For these values of μ, determine a basis of $\text{Ker } f$.

6.31 For the linear mapping $f : \mathbb{R}^3 \to \mathbb{R}^3$ given by

$$f(x, y, z) = (x + y, \, 0, \, y - z)$$

determine $\text{Im } f$, $\text{Ker } f$, and a basis of each.

If A is the subspace $\{(x, y, z) \in \mathbb{R}^3 \; ; \; x = y\}$, determine $f^{\leftarrow}(A)$ and a basis of it.

6.32 Let $f : \mathbb{R}^3 \to \mathbb{R}_3[X]$ be the linear mapping such that

$$f(1,0,0) = 2X + X^3;$$
$$f(0,1,0) = -2X + X^2;$$
$$f(0,0,1) = X^2 + X^3.$$

Determine

(1) $f(x,y,z)$ for all $(x,y,z) \in \mathbb{R}^3$;

(2) Im f and a basis of it;

(3) Ker f and a basis of it.

Extend the basis of (3) to a basis of $\mathbb{R}^3$.

6.33 Show that $\{a,b,c\}$ is a basis of $\mathbb{R}^3$ where

$$a = (-1,1,1), \quad b = (1,-1,1), \quad c = (1,1,-1).$$

Let $f : \mathbb{R}^3 \to \mathbb{R}^4$ be the linear mapping such that

$$f(a) = (1,0,1,\lambda), \quad f(b) = (0,1,-1,0), \quad f(c) = (1,-1,\lambda,-1).$$

(1) Determine $f(x,y,z)$ for all $(x,y,z) \in \mathbb{R}^3$.

(2) For which values of λ is f injective?

(3) Consider the subspace W of $\mathbb{R}^4$ given by $W = \text{Span}\,\{f(a),f(b)\}$. Determine dim W when $\lambda = -1$.

(4) With $\lambda = 2$ determine $f(1,1,0)$ and $f^{\leftarrow}\{(1,1,0,0)\}$.

6.34 A non-empty subset S of a vector space is **convex** if $tx + (1-t)y \in S$ for all $x,y \in S$ and all $t \in [0,1]$.

Prove that if S is a convex subset of $\mathbb{R}^n$ and $f : \mathbb{R}^n \to \mathbb{R}^n$ is linear then $f^{\to}(S)$ is also convex.

6.35 A diagram of finite-dimensional vector spaces and linear mappings of the form

$$V_1 \xrightarrow{f_1} V_2 \xrightarrow{f_2} V_3 \xrightarrow{f_3} \cdots \xrightarrow{f_n} V_{n+1}$$

is called an **exact sequence** if

(1) f_1 is injective;

(2) f_n is surjective;

(3) $(i = 1, \ldots, n-1)$ Im f_i = Ker f_{i+1}.

Prove that, for such an exact sequence,

$$\sum_{i=1}^{n+1}(-1)^i \dim V_i = 0.$$

6.36 Determine the rank and nullity of the linear mapping

$$f : \text{Mat}_{3\times 1}\,\mathbb{R} \to \text{Mat}_{3\times 1}\,\mathbb{R}$$

given by $f(\mathbf{x}) = A\mathbf{x}$ where

$$A = \begin{bmatrix} 1 & 0 & 2 \\ 2 & 1 & 2 \\ 0 & 1 & -2 \end{bmatrix}.$$

6.37 If $f : V \to W$ and $g : W \to X$ are linear, prove that

(1) $\operatorname{Im}(g \circ f) \subseteq \operatorname{Im} g$;

(2) $\operatorname{Ker} f \subseteq \operatorname{Ker}(g \circ f)$;

(3) $\operatorname{rank} f + \operatorname{rank} g - \dim W \leqslant \operatorname{rank} g \circ f \leqslant \min\{\operatorname{rank} f, \operatorname{rank} g\}$.

6.38 Determine the rank and nullity of the linear mapping

$$f : \mathbb{R}_3[X] \to \mathbb{R}_3[X]$$

given by

$$f(p(X)) = (X - 1)D^3 p(X).$$

6.39 Given $x = (x_1, x_2, x_3)$ and $y = (y_1, y_2, y_3)$ in $\mathbb{R}^3$, define the **wedge product** of x, y by

$$x \wedge y = (x_2 y_3 - x_3 y_2, \ x_3 y_1 - x_1 y_3, \ x_1 y_2 - x_2 y_1).$$

Define $f_y : \mathbb{R}^3 \to \mathbb{R}^3$ by

$$f_y(x) = x \wedge y.$$

Show that f_y is linear. If $y \neq 0$, prove that $\operatorname{Ker} f_y$ is the subspace spanned by $\{y\}$.

6.40 Let V be the real vector space of 2×2 hermitian matrices. Prove that the mapping $f : \mathbb{R}^4 \to V$ given by

$$f(x, y, z, w) = \begin{bmatrix} w + x & y + iz \\ y - iz & w - x \end{bmatrix}$$

is an isomorphism.

7.
The Matrix Connection

We shall now proceed to show how a linear mapping from one finite-dimensional vector space to another can be represented by a matrix. For this purpose, we require the following notion.

Definition

Let V be a finite-dimensional vector space over a field F. By an **ordered basis** of V we shall mean a finite sequence $(v_i)_{1 \leqslant i \leqslant n}$ of elements of V such that $\{v_1, \ldots, v_n\}$ is a basis of V.

Note that every basis of n elements gives rise to $n!$ distinct ordered bases, for there are $n!$ permutations on a set of n elements, and therefore $n!$ distinct ways of ordering the elements $v_1, \ldots, v_n$.

In what follows we shall find it convenient to abbreviate $(v_i)_{1 \leqslant i \leqslant n}$ to simply $(v_i)_n$.

Suppose now that V and W are vector spaces of dimensions m and n respectively over a field F. Let $(v_i)_m$, $(w_i)_n$ be given ordered bases of V, W and let $f : V \to W$ be linear. We know from Corollary 1 of Theorem 6.7 that f is completely and uniquely determined by its action on the basis $(v_i)_m$. This action is described by expressing each $f(v_i)$ as a linear combination of elements from the basis $(w_i)_n$:

$$f(v_1) = x_{11}w_1 + x_{12}w_2 + \cdots + x_{1n}w_n;$$
$$f(v_2) = x_{21}w_1 + x_{22}w_2 + \cdots + x_{2n}w_n;$$
$$\vdots$$
$$f(v_m) = x_{m1}w_1 + x_{m2}w_2 + \cdots + x_{mn}w_n.$$

The action of f on $(v_i)_m$ is therefore determined by the mn scalars x_{ij} appearing in the above equations. Put another way, *the action of f is completely determined by a knowledge of the $m \times n$ matrix $X = [x_{ij}]$.*

For technical reasons that will be explained later, the *transpose* of this matrix X

is called the **matrix of f relative to the fixed ordered bases** $(v_i)_m, (w_i)_n$. When it is clear what these fixed ordered bases are, we denote the matrix in question by Mat f.

- The reader should note carefully that it is an **n** × **m** matrix that represents a linear mapping from an **m**-dimensional vector space to an **n**-dimensional vector space.

Example 7.1

Consider the linear mapping $f : \mathbb{R}^3 \to \mathbb{R}^2$ given by

$$f(x, y, z) = (2x - 3y + z, \ 3x - 2y).$$

The action of f on the natural basis of $\mathbb{R}^3$ is described in terms of the natural basis of $\mathbb{R}^2$ as follows :

$$
\begin{aligned}
f(1,0,0) &= \quad (2,3) = \quad 2(1,0) + 3(0,1) \\
f(0,1,0) &= (-3,-2) = -3(1,0) - 2(0,1) \\
f(0,0,1) &= \quad (1,0) = \quad 1(1,0) + 0(0,1)
\end{aligned}
$$

and so we see that the matrix of f relative to the natural ordered bases of $\mathbb{R}^3$ and $\mathbb{R}^2$ is the transpose of the above coefficient matrix, namely the 2×3 matrix

$$
\begin{bmatrix}
2 & -3 & 1 \\
3 & -2 & 0
\end{bmatrix}.
$$

- Note how the rows of this matrix relate to the definition of f.

Example 7.2

The vector space $\mathbb{R}_n[X]$ is of dimension $n + 1$ and has the natural ordered basis

$$\{1, X, X^2, \dots, X^n\}.$$

The differentiation mapping $D : \mathbb{R}_n[X] \to \mathbb{R}_n[X]$ is linear, and

$$
\begin{aligned}
D1 \ &= 0 \cdot 1 + 0 \cdot X + \cdots + 0 \cdot X^{n-1} + 0 \cdot X^n \\
DX \ &= 1 \cdot 1 + 0 \cdot X + \cdots + 0 \cdot X^{n-1} + 0 \cdot X^n \\
DX^2 &= 0 \cdot 1 + 2 \cdot X + \cdots + 0 \cdot X^{n-1} + 0 \cdot X^n \\
&\ \ \vdots \\
DX^n &= 0 \cdot 1 + 0 \cdot X + \cdots + n \cdot X^{n-1} + 0 \cdot X^n
\end{aligned}
$$

so the matrix of D relative to the natural ordered basis of $\mathbb{R}_n[X]$ is the $(n+1) \times (n+1)$ matrix

$$
\begin{bmatrix}
0 & 1 & 0 & \cdots & 0 \\
0 & 0 & 2 & \cdots & 0 \\
\vdots & \vdots & \vdots & \ddots & \vdots \\
0 & 0 & 0 & \cdots & n \\
0 & 0 & 0 & \cdots & 0
\end{bmatrix}.
$$

EXERCISES

7.1 Consider the linear mapping $f : \mathbb{R}^2 \to \mathbb{R}^3$ given by

$$f(x, y) = (x + 2y, \ 2x - y, \ -x).$$

Determine the matrix of f
(1) relative to the natural ordered bases;
(2) relative to the ordered bases

$$\{(0, 1), (1, 1)\} \quad \text{and} \quad \{(0, 0, 1), (0, 1, 1), (1, 1, 1)\}.$$

7.2 Consider the linear mapping $f : \mathbb{R}^3 \to \mathbb{R}^2$ given by

$$f(x, y, z) = (2x - y, \ 2y - z).$$

Determine the matrix of f
(1) relative to the natural ordered bases;
(2) relative to the ordered bases

$$\{(1, 1, 1), (0, 1, 1), (0, 0, 1)\} \quad \text{and} \quad \{(0, 1), (1, 1)\}.$$

7.3 Consider the linear mapping $f : \mathbb{R}^3 \to \mathbb{R}^3$ given by

$$f(x, y, z) = (2x + z, \ y - x + z, \ 3z).$$

Determine the matrix of f with respect to the ordered basis

$$\{(1, -1, 0), (1, 0, -1), (1, 0, 0)\}.$$

7.4 Suppose that the mapping $f : \mathbb{R}^3 \to \mathbb{R}^3$ is linear and such that

$$f(1, 0, 0) = (2, 3, -2), \ f(1, 1, 0) = (4, 1, 4), \ f(1, 1, 1) = (5, 1, -7).$$

Find the matrix of f relative to the natural ordered basis of $\mathbb{R}^3$.

It is natural to ask what are the matrices that represent sums and scalar multiples of linear mappings. The answer is as follows.

Theorem 7.1

If V, W are of dimensions m, n respectively and if $f, g : V \to W$ are linear then, relative to fixed ordered bases,

$$\text{Mat}\,(f + g) = \text{Mat}\,f + \text{Mat}\,g$$

and, for every scalar λ,

$$\text{Mat}\,\lambda f = \lambda\,\text{Mat}\,f.$$

Proof

Let Mat $f = [x_{ij}]_{n \times m}$ and Mat $g = [y_{ij}]_{n \times m}$ relative to fixed ordered bases $(v_i)_m$ of V and $(w_i)_n$ of W. Then for $i = 1, \ldots, m$ we have (recalling the transposition involved)

$$f(v_i) = \sum_{j=1}^{n} x_{ji} w_j, \qquad g(v_i) = \sum_{j=1}^{n} y_{ji} w_j.$$

It follows that

$$(f + g)(v_i) = \sum_{j=1}^{n} (x_{ji} + y_{ji}) w_j$$

and therefore Mat $(f + g) = [x_{ij} + y_{ij}]_{n \times m} = $ Mat $f + $ Mat g.

Similarly, for every scalar λ we have

$$(\lambda f)(v_i) = \sum_{j=1}^{n} \lambda x_{ji} w_j$$

and so Mat $\lambda f = \lambda$ Mat f. $\square$

We can express Theorem 7.1 in a neater way as follows.

Consider the set $\text{Lin}_{m,n}(V, W)$ of linear mappings from a vector space V of dimension m to a vector space W of dimension n (each over the same field F). It is clear that, under the usual addition and multiplication by scalars, $\text{Lin}_{m,n}(V, W)$ is a vector space. Consider now the mapping

$$\vartheta : \text{Lin}_{m,n}(V, W) \rightarrow \text{Mat}_{n \times m} F$$

given by

$$\vartheta(f) = \text{Mat } f$$

where V and W are referred to fixed ordered bases $(v_i)_m$ and $(w_i)_n$ throughout.

This mapping ϑ is surjective. To see this, observe that given any $n \times m$ matrix $M = [m_{ij}]$, we can define

$$(i = 1, \ldots, m) \qquad f(v_i) = \sum_{j=1}^{n} m_{ji} w_j.$$

By Theorem 6.7, this produces a linear mapping $f : V \rightarrow W$; and clearly we have Mat $f = M$.

Moreover, ϑ is injective. This follows immediately from Corollary 2 of Theorem 6.7 and the definition of the matrix of a linear mapping.

Thus ϑ is a bijection and, by Theorem 7.1, is such that $\vartheta(f + g) = \vartheta(f) + \vartheta(g)$ and $\vartheta(\lambda f) = \lambda \vartheta(f)$. In other words, ϑ is a vector space isomorphism and we have:

Theorem 7.2

If V, W are of dimensions m, n respectively over F then

$$\text{Lin}_{m,n}(V, W) \simeq \text{Mat}_{n \times m} F. \quad \square$$

It is reasonable to ask if, for given ordered bases $(v_i)_m$ of V and $(w_i)_n$ of W, there is a 'natural' basis for the vector space $\text{Lin}_{m,n}(V, W)$. Indeed there is, and this can be obtained from the natural basis of $\text{Mat}_{n \times m} F$, namely

$$\{E_{pq} \; ; \; p = 1, \ldots, n \quad \text{and} \quad q = 1, \ldots, m\}$$

where E_{pq} is the matrix that has 1 in the (p, q)-th position and 0 elsewhere.

To see this, consider the linear mapping $f_{pq} : V \to W$ given by

$$f_{pq}(v_i) = \begin{cases} w_p & \text{if } i = q; \\ 0 & \text{otherwise.} \end{cases}$$

Then we have

$$f_{pq}(v_1) = 0w_1 + \cdots + 0w_p + \cdots + 0w_n$$
$$\vdots$$
$$f_{pq}(v_q) = 0w_1 + \cdots + 1w_p + \cdots + 0w_n$$
$$\vdots$$
$$f_{pq}(v_m) = 0w_1 + \cdots + 0w_p + \cdots + 0w_n$$

from which we see that $\text{Mat}\, f_{pq} = E_{pq}$, i.e. that $\vartheta(f_{pq}) = E_{pq}$.

Now since the inverse of an isomorphism is also an isomorphism, it follows by Theorem 6.5 that a (natural) basis for $\text{Lin}_{m,n}(V, W)$ is

$$\{f_{pq} \; ; \; p = 1, \ldots, n \quad \text{and} \quad q = 1, \ldots, m\}.$$

We now turn our attention to the matrix that represents the composite of two linear mappings. It is precisely in investigating this that we shall see how the definition of a matrix product arises in a natural way, and why we have chosen to use the transpose in the definition of the matrix of a linear mapping.

Consider the following situation:

$$U; (u_i)_m \xrightarrow{\; f;A \;} V; (v_i)_n \xrightarrow{\; g;B \;} W; (w_i)_p$$

in which the notation $U; (u_i)_m$ for example denotes a vector space U with a fixed ordered basis $(u_i)_m$ and $f;A$ denotes a linear mapping f represented, relative to the fixed bases, by the matrix A.

The composite mapping is described by

$$U; (u_i)_m \xrightarrow{\; g \circ f \;} W; (w_i)_p.$$

What is the matrix of this composite linear mapping?

It is natural to expect that this will depend on A and B. That this is so is the substance of the following result.

Theorem 7.3

$\text{Mat}\,(g \circ f) = \text{Mat}\, g \cdot \text{Mat}\, f.$

Proof

To find Mat $(g \circ f)$ we must express each image $(g \circ f)(u_i)$ in terms of the ordered basis $(w_i)_p$ of W. Now since Mat $f = A$ we have

$$(i = 1, \ldots, m) \qquad f(u_i) = \sum_{j=1}^{n} a_{ji} v_j,$$

and since Mat $g = B$ we have

$$(j = 1, \ldots, n) \qquad g(v_j) = \sum_{k=1}^{p} b_{kj} w_k.$$

Thus, for each i,

$$g[f(u_i)] = g\Big(\sum_{j=1}^{n} a_{ji} v_j\Big) = \sum_{j=1}^{n} a_{ji} g(v_j)$$
$$= \sum_{j=1}^{n} a_{ji}\Big(\sum_{k=1}^{p} b_{kj} w_k\Big)$$
$$= \sum_{k=1}^{p}\Big(\sum_{j=1}^{n} b_{kj} a_{ji}\Big) w_k.$$

Consequently the (k, i)-th element of Mat $(g \circ f)$ is $\sum_{j=1}^{n} b_{kj} a_{ji}$, which is precisely the (k, i)-th element of $BA = $ Mat $g \cdot$ Mat f. $\quad\square$

Corollary

A square matrix is invertible if and only if it represents an isomorphism.

Proof

Suppose that A is an $n \times n$ matrix that is invertible. Then there is an $n \times n$ matrix B such that $BA = I_n$. Let V be a vector space of dimension n and let $(v_i)_n$ be a fixed ordered basis of V. If $f, g : V \to V$ are linear mappings that are represented by A, B respectively then by Theorem 7.3 we have that $g \circ f$ is represented by $BA = I_n$. It follows that $g \circ f = \mathrm{id}_V$ whence, by Theorem 6.5, f is an isomorphism.

Conversely, if $f : V \to V$ is an isomorphism that is represented by the matrix A then the existence of f^{-1} such that $f^{-1} \circ f = \mathrm{id}_V$ implies the existence of a matrix B (namely that representing f^{-1}) such that $BA = I_n$, whence A is invertible. $\quad\square$

Example 7.3

Consider $\mathbb{R}^3$ referred to the natural ordered basis. If we change reference to the ordered basis

$$\{(1, 1, 0), (1, 0, 1), (0, 1, 1)\}$$

then the matrix of the identity mapping is obtained from the equations

$$\text{id}(1,0,0) = (1,0,0) = \tfrac{1}{2}(1,1,0) + \tfrac{1}{2}(1,0,1) - \tfrac{1}{2}(0,1,1)$$
$$\text{id}(0,1,0) = (0,1,0) = \tfrac{1}{2}(1,1,0) - \tfrac{1}{2}(1,0,1) + \tfrac{1}{2}(0,1,1)$$
$$\text{id}(0,0,1) = (0,0,1) = -\tfrac{1}{2}(1,1,0) + \tfrac{1}{2}(1,0,1) + \tfrac{1}{2}(0,1,1)$$

i.e. it is

$$\frac{1}{2}\begin{bmatrix} 1 & 1 & -1 \\ 1 & -1 & 1 \\ -1 & 1 & 1 \end{bmatrix}.$$

The identity mapping being an isomorphism, this matrix is invertible.

- The reader should note that it is to maintain the same order in which g, f appear in Theorem 7.3 that we choose to call the *transpose* of the coefficient matrix the matrix of the linear mapping. If, as some authors do, we were to write mappings on the right (i.e. write xf instead of $f(x)$) then this convention is unnecessary.

EXERCISES

7.5 A linear mapping $f : \mathbb{R}^3 \to \mathbb{R}^3$ is such that

$$f(1,0,0) = (0,0,1), \ f(1,1,0) = (0,1,1), \ f(1,1,1) = (1,1,1).$$

Determine $f(x,y,z)$ for all $(x,y,z) \in \mathbb{R}^3$ and compute the matrix of f relative to the ordered basis

$$B = \{(1,2,0),(2,1,0),(0,2,1)\}.$$

If $g : \mathbb{R}^3 \to \mathbb{R}^3$ is the linear mapping given by

$$g(x,y,z) = (2x, \ y+z, \ -x),$$

compute the matrix of $f \circ g \circ f$ relative to the ordered basis B.

7.6 Show that the matrix

$$A = \begin{bmatrix} 2 & 0 & 4 \\ -1 & 1 & -2 \\ 2 & 3 & 3 \end{bmatrix}$$

is invertible. If $f : \mathbb{R}^3 \to \mathbb{R}^3$ is a linear mapping whose matrix relative to the natural ordered basis of $\mathbb{R}^3$ is A, determine the matrix of f^{-1} relative to the same ordered basis.

7.7 If a linear mapping $f : V \to V$ is represented by the matrix A prove that f^n is represented by A^n.

We now consider the following important question. Suppose that we have the situation

$$V; (v_i)_m \xrightarrow{\ f;A\ } W; (w_i)_n.$$

If we refer V to a new ordered basis $(v_i')_m$ and W to a new ordered basis $(w_i')_n$ then clearly the matrix of f will change. How does it change?

To see how we can proceed, consider the particular case where $W = V$ and f is the identity mapping on V. We then have the situation

$$V; (v_i)_m \xrightarrow{\ \mathrm{id}_V;A\ } V; (v_i')_m$$
$$\uparrow \qquad\qquad\qquad\qquad \uparrow$$
$$\text{old basis} \qquad\qquad\qquad \text{new basis}$$

- This is precisely the situation described in the previous example.

We call A the **transition matrix from the basis** $(v_i)_m$ **to the basis** $(v_i')_m$.

The following result is clear from the Corollary of Theorem 7.3:

Theorem 7.4

Transition matrices are invertible. □

We can now describe how a change of bases is governed by the transition matrices that are involved.

Theorem 7.5

[Change of bases] *If a linear mapping $f : V \to W$ is represented relative to ordered bases $(v_i)_m$, $(w_i)_n$ by the $n \times m$ matrix A then relative to new ordered bases $(v_i')_m$, $(w_i')_n$ the matrix representing f is the $n \times m$ matrix $Q^{-1}AP$ where Q is the transition matrix from $(w_i')_n$ to $(w_i)_n$ and P is the transition matrix from $(v_i')_m$ to $(v_i)_m$.*

Proof

Using the notation introduced above, consider the diagram

$$
\begin{array}{ccc}
V; (v_i)_m & \xrightarrow{\ f;A\ } & W; (w_i)_n \\[2pt]
{\scriptstyle \mathrm{id}_V;P}\uparrow & & \uparrow{\scriptstyle \mathrm{id}_W;Q} \\[2pt]
V; (v_i')_m & \xrightarrow[\ f;X\ ]{} & W; (w_i')_n
\end{array}
$$

We have to determine the matrix X.

Now this diagram is 'commutative' in the sense that travelling from the south-west corner to the north-east corner is independent of whichever route we choose; for, clearly, $f \circ \mathrm{id}_V = f = \mathrm{id}_W \circ f$. It therefore follows by Theorem 7.3 that the matrices representing these routes are equal, i.e. that $AP = QX$. But Q, being a transition matrix, is invertible by Theorem 7.4 and so $X = Q^{-1}AP$. □

Example 7.4

Suppose that $f : \mathbb{R}^3 \to \mathbb{R}^3$ is the linear mapping whose matrix relative to the natural ordered basis is

$$A = \begin{bmatrix} 1 & -3 & 1 \\ 3 & -2 & 0 \\ -4 & 1 & 2 \end{bmatrix}$$

and let us compute the matrix of f when $\mathbb{R}^3$ is referred to the ordered basis

$$B = \{(1, -1, 1), (1, -2, 2), (1, -2, 1)\}.$$

We apply Theorem 7.5 with $W = V = \mathbb{R}^3$, $(w_i) = (v_i) =$ the natural ordered basis, and $(w'_i) = (v'_i) =$ the new ordered basis B.

The transition matrix from the new ordered basis to the old is

$$P = \begin{bmatrix} 1 & 1 & 1 \\ -1 & -2 & -2 \\ 1 & 2 & 1 \end{bmatrix}.$$

- Note that this is obtained by taking the elements of B and turning them into the columns of P. This becomes clear on observing that we have

$$\mathrm{id}(1, -1, 1) = (1, -1, 1) = 1(1, 0, 0) - 1(0, 1, 0) + 1(0, 0, 1)$$
$$\mathrm{id}(1, -2, 2) = (1, -2, 2) = 1(1, 0, 0) - 2(0, 1, 0) + 2(0, 0, 1)$$
$$\mathrm{id}(1, -2, 1) = (1, -2, 1) = 1(1, 0, 0) - 2(0, 1, 0) + 1(0, 0, 1)$$

and the transition matrix is the transpose of the coefficient matrix.

Now P is invertible (by Theorem 7.4) and the reader can easily verify that

$$P^{-1} = \begin{bmatrix} 2 & 1 & 0 \\ -1 & 0 & 1 \\ 0 & -1 & -1 \end{bmatrix}.$$

The matrix of f relative to the new ordered basis B is then, by Theorem 7.5,

$$P^{-1}AP = \begin{bmatrix} 15 & 25 & 23 \\ -8 & -11 & -12 \\ -2 & -5 & -3 \end{bmatrix}.$$

Example 7.5

Suppose that the linear mapping $f : \mathbb{R}^3 \to \mathbb{R}^2$ is represented, relative to the ordered bases $\{(1, 0, -1), (0, 2, 0), (1, 2, 3)\}$ of $\mathbb{R}^3$ and $\{(-1, 1), (2, 0)\}$ of $\mathbb{R}^2$, by the matrix

$$A = \begin{bmatrix} 2 & -1 & 3 \\ 3 & 1 & 0 \end{bmatrix}.$$

To determine the matrices that represent f relative to the natural ordered bases, we first determine the transition matrices P, Q from the natural ordered bases to the given ordered bases. Making use of the observation in the previous example, we can say immediately that

$$Q^{-1} = \begin{bmatrix} -1 & 2 \\ 1 & 0 \end{bmatrix}, \qquad P^{-1} = \begin{bmatrix} 1 & 0 & 1 \\ 0 & 2 & 2 \\ -1 & 0 & 3 \end{bmatrix}.$$

The reader can easily verify that

$$P = \tfrac{1}{4} \begin{bmatrix} 3 & 0 & -1 \\ -1 & 2 & -1 \\ 1 & 0 & 1 \end{bmatrix}$$

and hence that the required matrix is

$$Q^{-1}AP = \tfrac{1}{2} \begin{bmatrix} 3 & 3 & -5 \\ 5 & -1 & 1 \end{bmatrix}.$$

We now establish the converse of Theorem 7.5.

Theorem 7.6

Let $(v_i)_m, (w_i)_n$ be ordered bases of vector spaces V, W respectively. Suppose that A, B are $n \times m$ matrices such that there are invertible matrices P, Q such that $B = Q^{-1}AP$. Then there are ordered bases $(v_i')_m, (w_i')_n$ of V, W and a linear mapping $f : V \to W$ such that A is the matrix of f relative to $(v_i)_m, (w_i)_n$ and B is the matrix of f relative to $(v_i')_m, (w_i')_n$.

Proof

If $P = [p_{ij}]_{m \times m}$ and $Q = [q_{ij}]_{n \times n}$, define

$$(i = 1, \ldots, m) \quad v_i' = \sum_{j=1}^{m} p_{ji}v_j; \qquad (i = 1, \ldots, n) \quad w_i' = \sum_{j=1}^{n} q_{ji}w_j.$$

Since P is invertible there is, by the Corollary of Theorem 7.3, an isomorphism $f_P : V \to V$ that is represented by P relative to the ordered basis $(v_i)_m$. Since by definition $v_i' = f_P(v_i)$ for each i, it follows that $(v_i')_m$ is an ordered basis of V and that P is the transition matrix from $(v_i')_m$ to $(v_i)_m$. Similarly, $(w_i')_n$ is an ordered basis of W and Q is the transition matrix from $(w_i')_n$ to $(w_i)_n$.

Now let $f : V \to W$ be the linear mapping whose matrix, relative to the ordered bases $(v_i)_m$ and $(w_i)_n$ is A. Then by Theorem 7.5 the matrix of f relative to the ordered bases $(v_i')_m$ and $(w_i')_n$ is $Q^{-1}AP = B$. $\square$

EXERCISES

7.8 Determine the transition matrix from the ordered basis

$$\{(1,0,0,1), (0,0,0,1), (1,1,0,0), (0,1,1,0)\}$$

of $\mathbb{R}^4$ to the natural ordered basis of $\mathbb{R}^4$.

7.9 Consider the linear mapping $f : \mathbb{R}^3 \to \mathbb{R}^3$ given by

$$f(x, y, z) = (y, -x, z).$$

Compute the matrix A of f relative to the natural ordered basis and the B matrix of f relative to the ordered basis

$$\{(1, 1, 0), (0, 1, 1), (1, 0, 1)\}.$$

Determine an invertible matrix X such that $A = X^{-1}BX$.

7.10 Let $f : \mathbb{R}^3 \to \mathbb{R}^3$ be a linear mapping which is represented relative to the natural ordered basis by the matrix A. If P is the invertible matrix

$$P = \begin{bmatrix} 0 & 1 & 1 \\ 1 & 0 & 1 \\ 1 & 1 & 0 \end{bmatrix}$$

determine an ordered basis of $\mathbb{R}^3$ with respect to which the matrix of f is $P^{-1}AP$.

We have seen in Chapter 3 that a matrix of rank p can be transformed by means of row and column operations to the normal form

$$\begin{bmatrix} I_p & 0 \\ 0 & 0 \end{bmatrix}.$$

We can also deduce this as follows from the results we have established for linear mappings. The proof is of course more sophisticated.

Let V and W be of dimensions m and n respectively and let $f : V \to W$ be a linear mapping with dim Im $f = p$. By Theorem 6.4, we have

$$\dim \text{Ker } f = \dim V - \dim \text{Im } f = m - p,$$

so let $\{v_1, \ldots, v_{m-p}\}$ be a basis of Ker f. Using Corollary 1 of Theorem 5.8, extend this to a basis

$$B = \{u_1, \ldots, u_p, v_1, \ldots, v_{m-p}\}$$

of V. Observe now that

$$\{f(u_1), \ldots, f(u_p)\}$$

is linearly independent. In fact, if $\sum_{i=1}^{p} \lambda_i f(u_i) = 0$ then $f\left(\sum_{i=1}^{p} \lambda_i u_i\right) = 0$ and so

$\sum_{i=1}^{p} \lambda_i u_i \in \text{Ker } f$ whence $\sum_{i=1}^{p} \lambda_i u_i = \sum_{j=1}^{m-p} \mu_j v_j$. Then

$$\sum_{i=1}^{p} \lambda_i u_i - \sum_{j=1}^{m-p} \mu_j v_j = 0$$

and so, since B is a basis, every λ_i and every μ_j is 0.

It follows by Corollary 2 of Theorem 5.8 that $\{f(u_1), \ldots, f(u_p)\}$ is a basis of the subspace Im f. Now extend this, courtesy of Corollary 1 of Theorem 5.8, to a basis

$$C = \{f(u_1), \ldots, f(u_p), w_1, \ldots, w_{n-p}\}$$

of W. Then, since $f(v_1) = \cdots = f(v_{m-p}) = 0$, we have

$$f(u_1) = 1f(u_1) + 0f(u_2) + \cdots + 0f(u_p) + \cdots + 0w_{n-p};$$
$$f(u_2) = 0f(u_1) + 1f(u_2) + \cdots + 0f(u_p) + \cdots + 0w_{n-p};$$
$$\vdots$$
$$f(u_p) = 0f(u_1) + 0f(u_2) + \cdots + 1f(u_p) + \cdots + 0w_{n-p};$$
$$f(v_1) = 0f(u_1) + 0f(u_2) + \cdots + 0f(u_p) + \cdots + 0w_{n-p};$$
$$\vdots$$
$$f(v_{m-p}) = 0f(u_1) + 0f(u_2) + \cdots + 0f(u_p) + \cdots + 0w_{n-p}.$$

The matrix of f relative to the ordered bases B and C is then

$$\begin{bmatrix} I_p & 0 \\ 0 & 0 \end{bmatrix},$$

where p is the rank of f.

Suppose now that A is a given $n \times m$ matrix. If, relative to fixed ordered bases B_V, B_W this matrix represents the linear mapping $f : V \to W$ then, Q and P being the appropriate transition matrices from the bases B, C to the fixed ordered bases B_V and B_W, we have

$$Q^{-1}AP = \begin{bmatrix} I_p & 0 \\ 0 & 0 \end{bmatrix}.$$

Now since transition matrices are invertible they are products of elementary matrices. This means, therefore, that A can be reduced by means of row and column operations to the form

$$\begin{bmatrix} I_p & 0 \\ 0 & 0 \end{bmatrix}.$$

The above discussion shows, incidentally, that *the rank of a linear mapping f is the same as the rank of any matrix that represents f.*

Definition

If A, B are $n \times n$ matrices then B is said to be **similar** to A if there is an invertible matrix P such that $B = P^{-1}AP$.

It is clear that if B is similar to A then A is similar to B; for then

$$A = PBP^{-1} = (P^{-1})^{-1}AP^{-1}.$$

Also, if B is similar to A and C is similar to B then C is similar to A; for $B = P^{-1}AP$ and $C = Q^{-1}BQ$ give

$$C = Q^{-1}P^{-1}APQ = (PQ)^{-1}APQ.$$

Thus the relation of being similar is an equivalence relation on the set of $n \times n$ matrices.

The importance of similarity is reflected in the following result.

Theorem 7.6

Two $n \times n$ matrices A, B are similar if and only if they represent the same linear mapping relative to possibly different ordered bases.

Proof

This is immediate from Theorems 7.4 and 7.5 on taking $W = V$ and, for every i, $w_i = v_i$ and $w_i' = v_i'$. $\square$

Corollary

Similar matrices have the same rank. $\square$

The notion of similar matrices brings us back in a more concrete way to the discussion, at the end of Chapter 4, concerning the problem of deciding when (in our new terminology) a square matrix is similar to a diagonal matrix; or, equivalently, when a linear mapping can be represented by a diagonal matrix. We are not yet in a position to answer this question, but will proceed in the next chapter to develop some machinery that will help us towards this objective.

EXERCISES

7.11 Show that if matrices A, B are similar then so are A', B'.

7.12 Prove that if A, B are similar then so are A^k, B^k for all positive integers k.

7.13 Prove that, for every $\vartheta \in \mathbb{R}$, the complex matrices

$$\begin{bmatrix} \cos \vartheta & -\sin \vartheta \\ \sin \vartheta & \cos \vartheta \end{bmatrix}, \quad \begin{bmatrix} e^{i\vartheta} & 0 \\ 0 & e^{-i\vartheta} \end{bmatrix}$$

are similar.

SUPPLEMENTARY EXERCISES

7.14 Determine the matrix of the differentiation map D on $\mathbb{R}_n[X]$ relative to the ordered bases

(1) $\{1, X, X^2, \ldots, X^n\}$;

(2) $\{X^n, X^{n-1}, \ldots, X, 1\}$;

(3) $\{1, 1 + X, 1 + X^2, \ldots, 1 + X^n\}$.

7.15 Let V be a vector space of dimension n over a field F. A linear mapping $f : V \to V$ is said to be **nilpotent** if $f^p = 0$ for some positive integer p. The smallest such integer p is called the **index of nilpotency** of f. Suppose that f is nilpotent of index p. If $x \in V$ is such that $f^{p-1}(x) \neq 0$, prove that

$$\{x, f(x), f^2(x), \dots, f^{p-1}(x)\}$$

is linearly independent.

Hence prove that f is nilpotent of index n if and only if there is an ordered basis $(v_i)_n$ of V such that the matrix of f relative to $(v_i)_n$ is of the form

$$\begin{bmatrix} 0 & 0 & 0 & \dots & 0 & 0 \\ 1 & 0 & 0 & \dots & 0 & 0 \\ 0 & 1 & 0 & \dots & 0 & 0 \\ 0 & 0 & 1 & \dots & 0 & 0 \\ \vdots & \vdots & \vdots & & \vdots & \vdots \\ 0 & 0 & 0 & \dots & 1 & 0 \end{bmatrix}.$$

7.16 Let $f : \mathbb{R}_n[X] \to \mathbb{R}_n[X]$ be given by

$$f(p(X)) = p(X + 1).$$

Prove that f is linear and find the matrix of f relative to the natural ordered basis $\{1, X, \dots, X^n\}$.

7.17 Consider the mapping $f : \mathbb{R}_2[X] \to \text{Mat}_{2 \times 2} \mathbb{R}$ given by

$$f(a + bX + cX^2) = \begin{bmatrix} b + c & a \\ b & c \end{bmatrix}.$$

Show that f is linear and determine the matrix of f relative to the ordered bases $\{1, X, 1 + X^2\}$ of $\mathbb{R}_2[X]$ and

$$\left\{ \begin{bmatrix} 1 & 0 \\ 0 & 0 \end{bmatrix}, \begin{bmatrix} 1 & 1 \\ 0 & 0 \end{bmatrix}, \begin{bmatrix} 1 & 1 \\ 1 & 0 \end{bmatrix}, \begin{bmatrix} 1 & 1 \\ 1 & 1 \end{bmatrix} \right\}$$

of $\text{Mat}_{2 \times 2} \mathbb{R}$. Describe also $\text{Im } f$ and $\text{Ker } f$.

7.18 Let $f : \mathbb{R}_3[X] \to \mathbb{R}_3[X]$ be given by

$$f(a + bX + cX^2 + dX^3) = a + (d - c - a)X + (d - c)X^3.$$

Show that f is linear and determine the matrix of f relative to

(1) the natural ordered basis $\{1, X, X^2, X^3\}$;

(2) the ordered basis $\{1 + X^3, X, X + X^3, X^2 + X^3\}$.

7.19 If $A = [a_{ij}]_{n \times n}$ then the **trace** of A is defined to be the sum of the diagonal elements of A :

$$\text{tr } A = \sum_{i=1}^{n} a_{ii}.$$

Prove that

(1) $\mathrm{tr}(\lambda A) = \lambda\ \mathrm{tr}\ A$;

(2) $\mathrm{tr}(A + B) = \mathrm{tr}\ A + \mathrm{tr}\ B$;

(3) $\mathrm{tr}(AB) = \mathrm{tr}(BA)$.

Deduce from (3) that if A and B are similar then they have the same trace. Give an example of two matrices that have the same trace but are not similar.

8.
Determinants

In what follows it will be convenient to write an $n \times n$ matrix A in the form

$$A = [\mathbf{a}_1, \mathbf{a}_2, \ldots, \mathbf{a}_n]$$

where, as before, $\mathbf{a}_i$ represents the i-th column of A. Also, the letter F will signify either the field $\mathbb{R}$ of real numbers or the field $\mathbb{C}$ of complex numbers.

Definition

A mapping $D : \text{Mat}_{n \times n} F \to F$ is **determinantal** if it is

(a) **multilinear** (or *a linear function of each column*) in the sense that

(D_1) $D[\ldots, \mathbf{b}_i + \mathbf{c}_i, \ldots] = D[\ldots, \mathbf{b}_i, \ldots] + D[\ldots, \mathbf{c}_i, \ldots]$;
(D_2) $D[\ldots, \lambda \mathbf{a}_i, \ldots] = \lambda D[\ldots, \mathbf{a}_i, \ldots]$;

(b) **alternating** in the sense that

(D_3) $D[\ldots, \mathbf{a}_i, \ldots, \mathbf{a}_j, \ldots] = -D[\ldots, \mathbf{a}_j, \ldots, \mathbf{a}_i, \ldots]$;

(c) **1-preserving** in the sense that

(D_4) $D(I_n) = 1_F$.

We first observe that, in the presence of property (D_1), property (D_3) can be expressed in another way.

Theorem 8.1

If D satisfies property (D_1) then D satisfies property (D_3) if and only if it satisfies the property

(D_3') $D(A) = 0$ *whenever A has two identical columns.*

Proof

$\Rightarrow$: Suppose that A has two identical columns, say $\mathbf{a}_i = \mathbf{a}_j$ with $i \neq j$. Then by (D_3) we have $D(A) = -D(A)$ whence $D(A) = 0$.

$\Leftarrow$: Suppose now that D satisfies (D_1) and (D'_3). Then we have

$$0 \overset{(D'_3)}{=} D[\dots, \mathbf{a}_i + \mathbf{a}_j, \dots, \mathbf{a}_i + \mathbf{a}_j, \dots]$$
$$\overset{(D_1)}{=} D[\dots, \mathbf{a}_i, \dots, \mathbf{a}_i + \mathbf{a}_j, \dots] + D[\dots, \mathbf{a}_j, \dots, \mathbf{a}_i + \mathbf{a}_j, \dots]$$
$$\overset{(D_1)}{=} D[\dots, \mathbf{a}_i, \dots, \mathbf{a}_i, \dots] + D[\dots, \mathbf{a}_i, \dots, \mathbf{a}_j, \dots]$$
$$\qquad + D[\dots, \mathbf{a}_j, \dots, \mathbf{a}_i, \dots] + D[\dots, \mathbf{a}_j, \dots, \mathbf{a}_j, \dots]$$
$$\overset{(D'_3)}{=} D[\dots, \mathbf{a}_i, \dots, \mathbf{a}_j, \dots] + D[\dots, \mathbf{a}_j, \dots, \mathbf{a}_i, \dots]$$

whence (D_3) follows. $\square$

Corollary

D is determinantal if and only if it satisfies $(D_1), (D_2), (D'_3), (D_4)$. $\square$

Example 8.1

Let $D : \mathrm{Mat}_{2 \times 2} F \to F$ be given by

$$D \begin{bmatrix} a_{11} & a_{12} \\ a_{21} & a_{22} \end{bmatrix} = a_{11} a_{22} - a_{12} a_{21}.$$

Then it is an easy exercise to show that D satisfies the properties $(D_1), (D_2), (D'_3), (D_4)$ and so is determinantal.

In fact, as we shall now show, *this is the only determinantal mapping definable on* $\mathrm{Mat}_{2 \times 2} F$.

For this purpose, let

$$\delta_1 = \begin{bmatrix} 1 \\ 0 \end{bmatrix}, \quad \delta_2 = \begin{bmatrix} 0 \\ 1 \end{bmatrix}$$

so that every $A \in \mathrm{Mat}_{2 \times 2} F$ can be written in the form

$$A = [a_{11}\delta_1 + a_{21}\delta_2, \; a_{12}\delta_1 + a_{22}\delta_2].$$

Suppose that $f : \mathrm{Mat}_{2 \times 2} F \to F$ is determinantal. Then, by (D_1) we have

$$f(A) = f[a_{11}\delta_1, \; a_{12}\delta_1 + a_{22}\delta_2] + f[a_{21}\delta_2, \; a_{12}\delta_1 + a_{22}\delta_2].$$

Applying (D_1) again, the first summand can be expanded to

$$f[a_{11}\delta_1, \; a_{12}\delta_1] + f[a_{11}\delta_1, \; a_{22}\delta_2]$$

which, by (D_2), is

$$a_{11} a_{12} f[\delta_1, \delta_1] + a_{11} a_{22} f[\delta_1, \delta_2].$$

By (D'_3) and (D_4), this reduces to $a_{11} a_{22}$.

As for the second summand, by (D_1) this can be expanded to

$$f[a_{21}\delta_2, \; a_{12}\delta_1] + f[a_{21}\delta_2, \; a_{22}\delta_2]$$

which, by (D_2), is

$$a_{21}a_{12}f[\delta_2, \delta_1] + a_{21}a_{22}f[\delta_2, \delta_2].$$

By (D_3') and (D_4), this reduces to $-a_{21}a_{12}$.

Thus $f(A) = a_{11}a_{22} - a_{12}a_{21}$ and we conclude that there is a unique determinantal mapping on $\text{Mat}_{2\times 2} F$.

In what follows our objective will be to extend the above observation to $\text{Mat}_{n\times n} F$ for every positive integer n. The case where $n = 1$ is of course trivial, for if A is the 1×1 matrix $[a]$ then clearly the only determinantal mapping is that given by $D(A) = a$.

For every $A \in \text{Mat}_{n\times n} F$ we denote by A_{ij} the $(n-1) \times (n-1)$ matrix obtained from A by deleting the i-th row and the j-th column of A (i.e. the row and column containing a_{ij}).

The following result shows how we can construct determinantal mappings on the set of $n \times n$ matrices from a given determinantal mapping on the set of $(n-1) \times (n-1)$ matrices.

Theorem 8.2

For $n \geqslant 3$ let $D : \text{Mat}_{(n-1)\times(n-1)} F \to F$ be determinantal, and for $i = 1, \ldots, n$ define $f_i : \text{Mat}_{n\times n} F \to F$ by

$$f_i(A) = \sum_{j=1}^{n}(-1)^{i+j}a_{ij}D(A_{ij}).$$

Then each f_i is determinantal.

Proof

It is clear that $D(A_{ij})$ is independent of the j-th column of A and so $a_{ij}D(A_{ij})$ depends linearly on the j-th column of A. Consequently, we see that f_i depends linearly on the columns of A, i.e. that the properties (D_1) and (D_2) hold for f_i.

We now show that f_i satisfies (D_3'). For this purpose, suppose that A has two identical columns, say the p-th and the q-th columns with $p \neq q$. Then for $j \neq p$ and $j \neq q$ the $(n-1) \times (n-1)$ matrix A_{ij} has two identical columns and so, since D is determinantal by hypothesis, we have

$$(j \neq p, q) \qquad D(A_{ij}) = 0.$$

It follows that the above expression for $f_i(A)$ reduces to

$$f_i(A) = (-1)^{i+p}a_{ip}D(A_{ip}) + (-1)^{i+q}a_{iq}D(A_{iq}).$$

Suppose, without loss of generality, that $p < q$. Then it is clear that A_{iq} can be transformed into A_{ip} by effecting $q - 1 - p$ interchanges of adjacent columns; so, by (D_3) for D, we have

$$D(A_{iq}) = (-1)^{q-1-p}D(A_{ip}).$$

Since $a_{ip} = a_{iq}$ by hypothesis, we thus have

$$f_i(A) = [(-1)^{i+p} + (-1)^{i+q}(-1)^{q-1-p}]a_{ip}D(A_{ip})$$

which reduces to 0 since

$$(-1)^{i+p} + (-1)^{i+q}(-1)^{q-1-p} = (-1)^{i+p}[1 + (-1)^{2q-2p-1}]$$
$$= (-1)^{i+p}[1 + (-1)]$$
$$= 0.$$

Finally, f_i satisfies (D$_4$) since if $A = I_n$ then $a_{ij} = \delta_{ij}$ and $A_{ii} = I_{n-1}$, so that

$$f_i(I_n) = (-1)^{1+i}\delta_{ii}D(I_{n-1}) = 1.$$

Thus f_i is determinantal for every i. $\square$

Corollary

For every positive integer n there is at least one determinantal mapping on Mat $_{n \times n}$ *F*.

Proof

We proceed by induction. By Example 8.1, the result is true for $n = 2$. The inductive step is Theorem 8.2 which shows that a determinantal mapping can be defined on Mat $_{n \times n}$ F from a given determinantal mapping on Mat $_{(n-1) \times (n-1)}$ F. $\square$

Example 8.2

If D is the determinantal mapping on the set Mat $_{2 \times 2}$ F, i.e. if

$$D\begin{bmatrix} a_{11} & a_{12} \\ a_{21} & a_{22} \end{bmatrix} = a_{11}a_{22} - a_{12}a_{21},$$

then by Theorem 8.2 the mapping $f_1 : $ Mat $_{3 \times 3}$ $F \to F$ given by

$$f_1\begin{bmatrix} a_{11} & a_{12} & a_{13} \\ a_{21} & a_{22} & a_{23} \\ a_{31} & a_{32} & a_{33} \end{bmatrix}$$

$$= a_{11}D\begin{bmatrix} a_{22} & a_{23} \\ a_{32} & a_{33} \end{bmatrix} - a_{12}D\begin{bmatrix} a_{21} & a_{23} \\ a_{31} & a_{33} \end{bmatrix} + a_{13}D\begin{bmatrix} a_{21} & a_{22} \\ a_{31} & a_{32} \end{bmatrix}$$

is determinantal. Likewise, so are f_2 and f_3 given by

$$f_2\begin{bmatrix} a_{11} & a_{12} & a_{13} \\ a_{21} & a_{22} & a_{23} \\ a_{31} & a_{32} & a_{33} \end{bmatrix}$$

$$= a_{21}D\begin{bmatrix} a_{12} & a_{13} \\ a_{32} & a_{33} \end{bmatrix} - a_{22}D\begin{bmatrix} a_{11} & a_{13} \\ a_{31} & a_{33} \end{bmatrix} + a_{23}D\begin{bmatrix} a_{11} & a_{12} \\ a_{31} & a_{32} \end{bmatrix},$$

$$f_3 \begin{bmatrix} a_{11} & a_{12} & a_{13} \\ a_{21} & a_{22} & a_{23} \\ a_{31} & a_{32} & a_{33} \end{bmatrix}$$

$$= a_{31} D \begin{bmatrix} a_{12} & a_{13} \\ a_{22} & a_{23} \end{bmatrix} - a_{32} D \begin{bmatrix} a_{11} & a_{13} \\ a_{21} & a_{23} \end{bmatrix} + a_{33} D \begin{bmatrix} a_{11} & a_{12} \\ a_{21} & a_{22} \end{bmatrix}.$$

EXERCISES

8.1 Using the formula

$$D \begin{bmatrix} a_{11} & a_{12} \\ a_{21} & a_{22} \end{bmatrix} = a_{11} a_{22} - a_{12} a_{21},$$

evaluate each of $f_1(A), f_2(A), f_3(A)$ in the above example. What do you observe?

8.2 Show that the mapping $f : \text{Mat}_{3 \times 3} F \to F$ given by

$$f \begin{bmatrix} a_{11} & a_{12} & a_{13} \\ a_{21} & a_{22} & a_{23} \\ a_{31} & a_{32} & a_{33} \end{bmatrix}$$

$$= a_{11} D \begin{bmatrix} a_{22} & a_{23} \\ a_{32} & a_{33} \end{bmatrix} - a_{21} D \begin{bmatrix} a_{12} & a_{13} \\ a_{32} & a_{33} \end{bmatrix} + a_{31} D \begin{bmatrix} a_{12} & a_{13} \\ a_{22} & a_{23} \end{bmatrix}$$

is determinantal.

Our objective now is to establish the uniqueness of a determinantal mapping on $\text{Mat}_{n \times n} F$ for every positive integer n. For this purpose, it is necessary to digress a little and consider certain properties of permutations (= bijections) on a finite set.

On the set $\{1, \dots, n\}$ it is useful to write a permutation f in the form

$$\begin{pmatrix} 1 & 2 & 3 & \dots & n \\ f(1) & f(2) & f(3) & \dots & f(n) \end{pmatrix}.$$

Example 8.3

The permutation f on $\{0, \dots, 9\}$ described by $f(x) = x+1$ modulo 9 can be described by

$$f = \begin{pmatrix} 0 \ 1 \ 2 \ 3 \ 4 \ 5 \ 6 \ 7 \ 8 \ 9 \\ 1 \ 2 \ 3 \ 4 \ 5 \ 6 \ 7 \ 8 \ 9 \ 0 \end{pmatrix}.$$

Example 8.4

The permutation on $\{1, 2, 3, 4, 5\}$ described by

$$f = \begin{pmatrix} 1 \ 2 \ 3 \ 4 \ 5 \\ 1 \ 2 \ 4 \ 5 \ 3 \end{pmatrix}$$

'fixes' 1 and 2, and permutes cyclically $3, 4, 5$.

Given permutations f, g on $\{1, \ldots, n\}$ we can compute the composite permutation $g \circ f$ by simply treating them as mappings:

$$\begin{pmatrix} 1 & \ldots & n \\ g(1) & \ldots & g(n) \end{pmatrix} \circ \begin{pmatrix} 1 & \ldots & n \\ f(1) & \ldots & f(n) \end{pmatrix} = \begin{pmatrix} 1 & \ldots & n \\ g[f(1)] & \ldots & g[f(n)] \end{pmatrix}.$$

Example 8.5

Consider the permutations

$$f = \begin{pmatrix} 1 \ 2 \ 3 \ 4 \ 5 \ 6 \\ 1 \ 6 \ 4 \ 3 \ 5 \ 2 \end{pmatrix}, \quad g = \begin{pmatrix} 1 \ 2 \ 3 \ 4 \ 5 \ 6 \\ 2 \ 6 \ 5 \ 3 \ 1 \ 4 \end{pmatrix}.$$

Working from the right, we compute the composite permutation $g \circ f$ as follows:

$$g \circ f = \begin{pmatrix} 1 \ 2 \ 3 \ 4 \ 5 \ 6 \\ 2 \ 6 \ 5 \ 3 \ 1 \ 4 \end{pmatrix} \circ \begin{pmatrix} 1 \ 2 \ 3 \ 4 \ 5 \ 6 \\ 1 \ 6 \ 4 \ 3 \ 5 \ 2 \end{pmatrix} = \begin{pmatrix} 1 \ 2 \ 3 \ 4 \ 5 \ 6 \\ 2 \ 4 \ 3 \ 5 \ 1 \ 6 \end{pmatrix}.$$

EXERCISES

8.3 Compute the products

$$\begin{pmatrix} 1 \ 2 \ 3 \ 4 \ 5 \ 6 \ 7 \ 8 \\ 8 \ 6 \ 2 \ 3 \ 1 \ 4 \ 5 \ 7 \end{pmatrix} \circ \begin{pmatrix} 1 \ 2 \ 3 \ 4 \ 5 \ 6 \ 7 \ 8 \\ 5 \ 6 \ 7 \ 1 \ 8 \ 4 \ 3 \ 2 \end{pmatrix};$$

$$\begin{pmatrix} 1 \ 2 \ 3 \ 4 \ 5 \ 6 \ 7 \ 8 \\ 8 \ 6 \ 2 \ 3 \ 1 \ 4 \ 5 \ 7 \end{pmatrix} \circ \begin{pmatrix} 1 \ 2 \ 3 \ 4 \ 5 \ 6 \ 7 \ 8 \\ 5 \ 3 \ 4 \ 6 \ 7 \ 2 \ 8 \ 1 \end{pmatrix}.$$

Definition

By a **transposition** on the set $\{1, 2, \ldots, n\}$ we mean a permutation that interchanges two elements and fixes the other elements. More precisely, a transposition is a permutation τ such that, for some i, j with $i \neq j$,

$$\tau(i) = j, \quad \tau(j) = i, \quad \text{and} \quad (\forall x \neq i, j) \ \tau(x) = x.$$

We shall sometimes denote the transposition τ that interchanges i and j by the notation

$$\tau : i \leftrightarrow j.$$

Clearly, the inverse of a transposition is also a transposition.

The set of permutations on $\{1, 2, \ldots, n\}$ will be denoted by P_n.

Theorem 8.3

If $n > 2$ then every $\sigma \in P_n$ can be expressed as a composite of transpositions.

Proof

We establish the result by induction on n. When $n = 2$ it is clear that σ is itself a transposition, in which case there is nothing to prove.

Suppose, by way of induction, that $n > 2$ and that the result holds for all permutations in P_{n-1}. Let $\sigma \in P_n$ and suppose that $\sigma(n) = k$. Let τ be the transposition

$$\tau : n \leftrightarrow k.$$

Then $\tau \circ \sigma$ is such that $(\tau \circ \sigma)(n) = n$; i.e. $\tau \circ \sigma$ fixes n, and so induces a permutation, $(\tau \circ \sigma)^\star$ say, in P_{n-1}. By the induction hypothesis, there are transpositions $\tau_1^\star, \ldots, \tau_r^\star$ in P_{n-1} such that $(\tau \circ \sigma)^\star = \tau_1^\star \circ \cdots \circ \tau_r^\star$. Clearly, $\tau_1^\star, \ldots, \tau_r^\star$ induce transpositions in P_n, say $\tau_1, \ldots, \tau_r$, each of which fixes n, and $\tau \circ \sigma = \tau_1 \circ \cdots \circ \tau_r$. It follows that

$$\sigma = \tau^{-1} \circ \tau_1 \circ \cdots \circ \tau_r$$

as required. $\square$

Given $\sigma \in P_n$ let $I(\sigma)$ be the number of **inversions** in σ, i.e. the number of pairs (i, j) with $i < j$ and $\sigma(j) < \sigma(i)$.

Definition

For every $\sigma \in P_n$ the **signum** (or **signature**) of σ is defined by $\varepsilon_\sigma = (-1)^{I(\sigma)}$.

Theorem 8.4

$(\forall \rho, \sigma \in P_n) \quad \varepsilon_{\rho\sigma} = \varepsilon_\rho \varepsilon_\sigma.$

Proof

Consider the product

$$V_n = \prod_{i<j} (j - i).$$

For every $\sigma \in P_n$ define

$$\sigma(V_n) = \prod_{i<j} [\sigma(j) - \sigma(i)].$$

Since σ is a bijection, every factor of V_n occurs precisely once in $\sigma(V_n)$, up to a possible change in sign. Consequently we have

$$\sigma(V_n) = (-1)^{I(\sigma)} V_n = \varepsilon_\sigma V_n.$$

Given $\rho, \sigma \in P_n$ we have similarly $\rho\sigma(V_n) = \varepsilon_\rho \sigma(V_n)$. Consequently,

$$\varepsilon_{\rho\sigma} V_n = \rho\sigma(V_n) = \varepsilon_\rho \sigma(V_n) = \varepsilon_\rho \varepsilon_\sigma V_n$$

whence, since $V_n \neq 0$, we obtain $\varepsilon_{\rho\sigma} = \varepsilon_\rho \varepsilon_\sigma$. $\square$

Corollary

If $\sigma \in P_n$ then $\varepsilon_\sigma = \pm 1$ and $\varepsilon_{\sigma^{-1}} = \varepsilon_\sigma$.

Proof

It is clear that $\varepsilon_\tau = -1$ for every transposition τ. It follows from Theorems 8.3 and 8.4 that $\varepsilon_\sigma = \pm 1$ for every permutation σ. Moreover, since the signum of the identity permutation is clearly 1, we deduce from $\varepsilon_\sigma \varepsilon_{\sigma^{-1}} = \varepsilon_{\sigma\sigma^{-1}} = 1$ that $\varepsilon_{\sigma^{-1}} = \varepsilon_\sigma$. $\square$

We say that σ is an **even** permutation if $\varepsilon_\sigma = 1$, and an **odd** permutation if $\varepsilon_\sigma = -1$. An even permutation is therefore one that has an even number of transpositions, whereas an odd permutation is one with an odd number of transpositions. This notion of parity is therefore an invariant associated with a permutation.

Example 8.6

Consider the permutation

$$f = \begin{pmatrix} 1\ 2\ 3\ 4\ 5\ 6 \\ 1\ 6\ 4\ 3\ 5\ 2 \end{pmatrix}.$$

If we join each i on the top line with the corresponding i on the bottom line we obtain the diagram

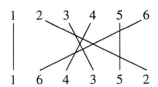

In this the number of distinct crossings gives the number of inversions. This is 8, which is even. The permutation is therefore even.

EXERCISES

8.4 Determine the parity of each of the following permutations:

$$\begin{pmatrix} 1\ 2\ 3\ 4\ 5\ 6 \\ 6\ 1\ 4\ 2\ 5\ 3 \end{pmatrix}, \quad \begin{pmatrix} 1\ 2\ 3\ 4\ 5\ 6 \\ 5\ 4\ 3\ 6\ 1\ 2 \end{pmatrix}.$$

Theorem 8.5

There is a unique determinantal map $D : \text{Mat}_{n \times n} F \to F$, and it can be described by

$$D(A) = \sum_{\sigma \in P_n} \varepsilon_\sigma a_{\sigma(1),1} \cdots a_{\sigma(n),n}.$$

Proof

We know by the Corollary of Theorem 8.2 that at least one determinantal mapping D exists on $\text{Mat}_{n \times n} F$. If we write δ_i for the i-th column of I_n then we can represent an $n \times n$ matrix $A = [a_{ij}]$ by

$$A = [a_{11}\delta_1 + \cdots + a_{n1}\delta_n, \ldots \ldots, a_{1n}\delta_1 + \cdots + a_{nn}\delta_n].$$

Using property (D_1) we can write $D(A)$ as a sum of terms of the form

$$D[a_{\sigma(1),1}\delta_{\sigma(1)}, \ldots, a_{\sigma(n),n}\delta_{\sigma(n)}],$$

where $1 \leqslant \sigma(i) \leqslant n$ for every i. Using property (D_2) we can then write each of these terms as

$$a_{\sigma(1),1} \cdots a_{\sigma(n),n} D[\delta_{\sigma(1)}, \ldots, \delta_{\sigma(n)}].$$

But, by property (D_3'), each such expression is 0 except those in which we have $\sigma(i) \neq \sigma(j)$ for $i \neq j$; i.e. those in which σ is a permutation on $\{1, \ldots, n\}$. Thus we have that

$$D(A) = \sum_{\sigma \in P_n} a_{\sigma(1),1} \cdots a_{\sigma(n),n} D[\delta_{\sigma(1)}, \ldots, \delta_{\sigma(n)}].$$

Now the columns $\delta_{\sigma(1)}, \ldots, \delta_{\sigma(n)}$ occur in the permutation σ of the standard arrangement $\delta_1, \ldots, \delta_n$. If, using Theorem 8.3, we write σ as a composite of transpositions, say

$$\sigma = \tau_1 \circ \cdots \circ \tau_k,$$

then we have

$$\sigma^{-1} = \tau_k^{-1} \circ \cdots \circ \tau_1^{-1}.$$

Restoring the standard arrangement of the columns by applying $\tau_1^{-1}, \ldots, \tau_k^{-1}$ we see, by property (D_3) and the fact that $\varepsilon_\sigma = \varepsilon_{\sigma^{-1}}$, that

$$D[\delta_{\sigma(1)}, \ldots, \delta_{\sigma(n)}] = \varepsilon_\sigma D[\delta_1, \ldots, \delta_n].$$

But by property (D_4) we have $D[\delta_1, \ldots, \delta_n] = D(I_n) = 1$. We therefore conclude that

$$D(A) = \sum_{\sigma \in P_n} \varepsilon_\sigma a_{\sigma(1),1} \cdots a_{\sigma(n),n}.$$

The above argument also shows that D is unique. $\square$

An important consequence of the above is that *the expression for $f_i(A)$ given in Theorem 8.2 is independent of i.*

Definition

The unique determinantal mapping on $\mathrm{Mat}_{n \times n} F$ will be denoted by $\det$. By the **determinant** of $A = [a_{ij}]_{n \times n}$ we shall mean $\det A$.

By Theorem 8.5, we have that

$$\det A = \sum_{\sigma \in P_n} \varepsilon_\sigma a_{\sigma(1),1} \cdots a_{\sigma(n),n}.$$

Alternatively, by Theorem 8.2, we have that, for $i = 1, \ldots, n$,

$$\det A = \sum_{j=1}^{n} (-1)^{i+j} a_{ij} \det A_{ij},$$

which is called the **Laplace expansion along the i-th row**.

- Note that, as pointed out above, the Laplace expansion is independent of the row chosen.

Example 8.7

Consider the matrix

$$A = \begin{bmatrix} 1 & 1 & -1 \\ 2 & 1 & 3 \\ 1 & -5 & 1 \end{bmatrix}.$$

Using a Laplace expansion along the first row, we have

$$\det A = 1 \cdot \det \begin{bmatrix} 1 & 3 \\ -5 & 1 \end{bmatrix} - 1 \cdot \det \begin{bmatrix} 2 & 3 \\ 1 & 1 \end{bmatrix} + (-1) \cdot \det \begin{bmatrix} 2 & 1 \\ 1 & -5 \end{bmatrix}$$
$$= 16 - (-1) - (-11) = 28.$$

Expanding along the second row, we obtain

$$\det A = -2 \cdot \det \begin{bmatrix} 1 & -1 \\ -5 & 1 \end{bmatrix} + 1 \cdot \det \begin{bmatrix} 1 & -1 \\ 1 & 1 \end{bmatrix} - 3 \cdot \det \begin{bmatrix} 1 & 1 \\ 1 & -5 \end{bmatrix}$$
$$= -2(-4) + 2 - 3(-6) = 28.$$

Finally, expanding along the third row we obtain

$$\det A = 1 \cdot \det \begin{bmatrix} 1 & -1 \\ 1 & 3 \end{bmatrix} - (-5) \cdot \det \begin{bmatrix} 1 & -1 \\ 2 & 3 \end{bmatrix} + 1 \cdot \det \begin{bmatrix} 1 & 1 \\ 2 & 1 \end{bmatrix}$$
$$= 4 + 5 \cdot 5 + (-1) = 28.$$

EXERCISES

8.5 Compute, via a third row Laplace expansion,

$$\det \begin{bmatrix} 1 & 2 & -3 & 4 \\ -4 & 2 & 1 & 3 \\ 3 & 0 & -2 & 0 \\ 1 & 0 & 2 & -5 \end{bmatrix}.$$

8.6 Determine

$$\det \begin{bmatrix} 1-\lambda & 3 & 2 \\ 2 & -1-\lambda & 3 \\ 3 & 0 & 1-\lambda \end{bmatrix}.$$

Theorem 8.6

If A is a square matrix then $\det A = \det A'$.

Proof

If $\sigma \in P_n$ then whenever $\sigma(i) = j$ we have $i = \sigma^{-1}(j)$ and therefore $a_{\sigma(i),i} = a_{j,\sigma^{-1}(j)}$. Consequently,

$$a_{\sigma(1),1} \cdots a_{\sigma(n),n} = a_{1,\sigma^{-1}(1)} \cdots a_{n,\sigma^{-1}(n)} = [A']_{\sigma^{-1}(1),1} \cdots [A']_{\sigma^{-1}(n),n}.$$

Now as σ ranges over P_n so does σ^{-1}; and $\varepsilon_\sigma = \varepsilon_{\sigma^{-1}}$. Thus

$$\det A = \sum_{\sigma \in P_n} \varepsilon_\sigma a_{\sigma(1),1} \cdots a_{\sigma(n),n} = \sum_{\sigma^{-1} \in P_n} \varepsilon_{\sigma^{-1}} [A']_{\sigma^{-1}(1),1} \cdots [A']_{\sigma^{-1}(n),n} = \det A'. \quad \square$$

Corollary

For $j = 1, \ldots, n$ we have

$$\det A = \sum_{i=1}^{n} (-1)^{i+j} a_{ij} \det A_{ij}.$$

Proof

We have

$$\det A = \det A' = \sum_{j=1}^{n} (-1)^{i+j} a_{ji} \det A_{ji}$$
$$= \sum_{i=1}^{n} (-1)^{i+j} a_{ij} \det A_{ij},$$

the second summation being obtained from the first summation by interchanging i and j. $\quad \square$

The reader should note that in the above Corollary the summation is over the *first* index whereas in Theorem 8.2 it is over the *second* index.

We can thus assert that *Laplace expansions via columns are also valid.*

EXERCISES

8.7 For the matrix A of the previous example, find $\det A$ by Laplace expansion via each of the three columns.

8.8 Compute

$$\det \begin{bmatrix} 1 & 2 & 3 & -4 \\ 0 & -5 & 6 & -7 \\ 0 & 0 & -8 & 9 \\ 0 & 0 & 0 & 10 \end{bmatrix}.$$

It is useful to know how row and column operations on a square matrix effect the determinant of that matrix. It is clear from property (D₃) that

- *if B is obtained from A by interchanging two columns of A then* det B = $-$ det A.

Also, from property (D₂) it follows that

- *if B is obtained from A by multiplying a column of A by some scalar* λ *then* det $B = \lambda$ det A.

Finally, if B is obtained from A by adding λ times the i-th column of A to the j-th column then, by properties (D₁) and (D′₃), we have

$$\det B = \det \left[\ldots, \mathbf{a}_i, \ldots, \mathbf{a}_j + \lambda \mathbf{a}_i, \ldots \right]$$
$$= \det \left[\ldots, \mathbf{a}_i, \ldots, \mathbf{a}_j, \ldots \right] + \det \left[\ldots, \mathbf{a}_i, \ldots, \lambda \mathbf{a}_i, \ldots \right]$$
$$= \det A + \lambda \det \left[\ldots, \mathbf{a}_i, \ldots, \mathbf{a}_i, \ldots \right]$$
$$= \det A + \lambda 0$$
$$= \det A.$$

Thus we have that

- *if B is obtained from A by adding to any column of A a multiple of another column then* det B = det A.

Since row operations on A are simply column operations on A^t, and since det A^t = det A, it is clear from the above that *similar observations hold for row operations.*

Example 8.8

Any square matrix that has a zero row or a zero column has a zero determinant. To see this, simply perform a Laplace expansion along that zero row or column.

Example 8.9

Consider again the matrix

$$A = \begin{bmatrix} 1 & 1 & -1 \\ 2 & 1 & 3 \\ 1 & -5 & 1 \end{bmatrix}.$$

Using row operations, we have

$$\det A = \det \begin{bmatrix} 1 & 1 & -1 \\ 0 & -1 & 5 \\ 0 & -6 & 2 \end{bmatrix} \quad \begin{matrix} \rho_2 - 2\rho_1 \\ \rho_3 - \rho_1 \end{matrix}$$
$$= \det \begin{bmatrix} -1 & 5 \\ -6 & 2 \end{bmatrix} \quad \text{by first column Laplace}$$
$$= -2 + 30 = 28.$$

Example 8.10

Consider the matrix

$$A = \begin{bmatrix} 1 & 0 & -1 \\ 2 & 0 & 3 \\ 1 & -5 & 1 \end{bmatrix}.$$

By a Laplace expansion using the second column, we have

$$\det A = -(-5)\det \begin{bmatrix} 1 & -1 \\ 2 & 3 \end{bmatrix} = 5 \cdot 5 = 25.$$

Example 8.11

For the matrix

$$A = \begin{bmatrix} 1 & -1 & 2 & 3 \\ 2 & 2 & 0 & 2 \\ 1 & 2 & 3 & 0 \\ 5 & 3 & 2 & -1 \end{bmatrix}$$

we have

$$
\begin{aligned}
\det A &= \begin{bmatrix} 1 & -1 & 2 & 3 \\ 0 & 4 & -4 & -4 \\ 0 & 3 & 1 & -3 \\ 0 & 8 & -8 & -16 \end{bmatrix} \quad \begin{matrix} \\ \rho_2 - 2\rho_1 \\ \rho_3 - \rho_1 \\ \rho_4 - 5\rho_1 \end{matrix} \\
&= \det \begin{bmatrix} 4 & -4 & -4 \\ 3 & 1 & -3 \\ 8 & -8 & -16 \end{bmatrix} \quad \text{by first column Laplace} \\
&= 4 \cdot 8 \cdot \det \begin{bmatrix} 1 & -1 & -1 \\ 3 & 1 & -3 \\ 1 & -1 & -2 \end{bmatrix} \\
&= 32 \det \begin{bmatrix} 1 & -1 & -1 \\ 0 & 4 & 0 \\ 0 & 0 & -1 \end{bmatrix} \quad \begin{matrix} \\ \rho_2 - 3\rho_1 \\ \rho_3 - \rho_1 \end{matrix} \\
&= 32(-4) = -128.
\end{aligned}
$$

EXERCISES

8.9 What is the determinant of an $n \times n$ elementary matrix?

8.10 Prove that the determinant of an $n \times n$ upper triangular matrix is the product of its diagonal elements.

8.11 Determine the values of x for which $\det A \neq 0$ where

$$A = \begin{bmatrix} x & 2 & 0 & 3 \\ 1 & 2 & 3 & 3 \\ 1 & 0 & 1 & 1 \\ 1 & 1 & 1 & 3 \end{bmatrix}.$$

8.12 Using row and column operations, show that

$$\det \begin{bmatrix} 0 & a & 0 & 0 & 0 & 0 \\ f & 0 & b & 0 & 0 & 0 \\ 0 & g & 0 & c & 0 & 0 \\ 0 & 0 & h & 0 & d & 0 \\ 0 & 0 & 0 & k & 0 & e \\ 0 & 0 & 0 & 0 & m & 0 \end{bmatrix} = -acefhm.$$

8.13 Compute the determinant of the matrix

$$\begin{bmatrix} 1 & 2 & 3 & \cdots & n \\ -1 & 0 & 3 & \cdots & n \\ -1 & -2 & 0 & \cdots & n \\ \vdots & \vdots & \vdots & \ddots & \vdots \\ -1 & -2 & -3 & \cdots & 0 \end{bmatrix}.$$

8.14 Compute the determinant of the matrix $A = [a_{ij}]_{n \times n}$ given by

$$a_{ij} = \begin{cases} 1 & \text{if } i + j = n + 1; \\ 0 & \text{otherwise.} \end{cases}$$

8.15 Consider the complex matrix

$$A = \begin{bmatrix} 0 & 1 + i & 1 + 2i \\ 1 - i & 0 & 2 - 3i \\ 1 - 2i & 2 + 3i & 0 \end{bmatrix}.$$

Show that $\det A = 6$.

We now consider some further important properties of determinants.

Theorem 8.7

If $A, B \in \text{Mat}_{n \times n} F$ then

$$\det AB = \det A \cdot \det B.$$

Proof

If $C = AB$ then the k-th column of C can be written

$$\mathbf{c}_k = b_{1k}\mathbf{a}_1 + \cdots + b_{nk}\mathbf{a}_n.$$

To see this, observe that the i-th element of $\mathbf{c}_k$ is

$$[\mathbf{c}_k]_i = c_{ik} = \sum_{j=1}^{n} a_{ij}b_{jk} = \sum_{j=1}^{n} b_{jk}[\mathbf{a}_j]_i = \left[\sum_{j=1}^{n} b_{jk}\mathbf{a}_j\right]_i.$$

Thus we have that

$$\det AB = \det C$$

$$= \det [b_{11}\mathbf{a}_1 + \cdots + b_{n1}\mathbf{a}_n, \ldots, b_{1n}\mathbf{a}_1 + \cdots + b_{nn}\mathbf{a}_n]$$

$$= \sum_{\sigma \in P_n} \det [b_{\sigma(1),1}\mathbf{a}_{\sigma(1)}, \ldots, b_{\sigma(n),n}\mathbf{a}_{\sigma(n)}]$$

$$= \sum_{\sigma \in P_n} b_{\sigma(1),1} \cdots b_{\sigma(n),n} \det [\mathbf{a}_{\sigma(1)}, \ldots, \mathbf{a}_{\sigma(n)}]$$

$$= \sum_{\sigma \in P_n} b_{\sigma(1),1} \cdots b_{\sigma(n),n} \epsilon_\sigma \det [\mathbf{a}_1, \ldots, \mathbf{a}_n]$$

$$= \det A \cdot \sum_{\sigma \in P_n} \epsilon_\sigma b_{\sigma(1),1} \cdots b_{\sigma(n),n}$$

$$= \det A \cdot \det B. \quad \square$$

Corollary

If A is invertible then $\det A \neq 0$ and $\det A^{-1} = \dfrac{1}{\det A}$.

Proof

This follows from $\det A \cdot \det A^{-1} = \det AA^{-1} = \det I_n = 1$. $\quad \square$

EXERCISES

8.16 Given the matrices

$$A = \begin{bmatrix} b + 8c & 2c - 2b & 4b - 4c \\ 4c - 4a & c + 8b & 2a - 2c \\ 2b - 2a & 4a - 4b & a + 8b \end{bmatrix}, \quad P = \begin{bmatrix} 0 & 1 & 2 \\ 2 & 0 & 1 \\ 1 & 2 & 0 \end{bmatrix}$$

find P^{-1} and compute $P^{-1}AP$. Hence determine $\det A$.

8.17 If A is a square matrix such that $A^p = 0$ for some positive integer p, prove that $\det A = 0$.

If A is an invertible $n \times n$ matrix then we have seen above that $\det A \neq 0$. Our objective now is to show that the converse of this holds. This will not only provide a useful way of determining when a matrix is invertible but will also give a new way of computing inverses. For this purpose, we require the following notion.

Definition

If $A \in \mathrm{Mat}_{n \times n} F$ then the **adjugate** (or **adjoint**) of A is the $n \times n$ matrix adj A given by

$$[\mathrm{adj}\, A]_{ij} = (-1)^{i+j} \det A_{ji}.$$

- It is important to note the *reversal of the suffices* in the above definition.

The adjugate matrix has the following useful property.

Theorem 8.8

For every $n \times n$ matrix A,

$$A \cdot \text{adj } A = (\det A)I_n = \text{adj } A \cdot A.$$

Proof

We have

$$
\begin{aligned}
[A \cdot \text{adj } A]_{ij} &= \sum_{k=1}^{n} a_{ik}[\text{adj } A]_{kj} \\
&= \sum_{k=1}^{n} a_{ik} \det A_{jk} \\
&= \begin{cases} \det A & \text{if } i = j; \\ 0 & \text{if } i \neq j, \end{cases}
\end{aligned}
$$

the last equality resulting from the fact that when $i \neq j$ the expression represents the determinant of a matrix whose j-th row is the same as its i-th row. Thus $A \cdot \text{adj } A$ is a diagonal matrix all of whose diagonal entries is $\det A$; in other words, $A \cdot \text{adj } A = (\det A)I_n$. The second equality is established similarly. □

EXERCISES

8.18 Compute the adjugate of each of the following matrices:

$$\begin{bmatrix} a & b \\ c & d \end{bmatrix}, \quad \begin{bmatrix} a & h & g \\ h & b & f \\ g & f & c \end{bmatrix}, \quad \begin{bmatrix} -1 & 0 & -1 \\ 0 & -1 & 0 \\ -1 & 0 & -1 \end{bmatrix}.$$

Theorem 8.9

A square matrix A is invertible if and only if $\det A \neq 0$, in which case the inverse is given by

$$A^{-1} = \frac{1}{\det A} \text{adj } A.$$

Proof

If $\det A \neq 0$ then by Theorem 8.8 we have

$$A \cdot \frac{1}{\det A} \text{adj } A = I_n$$

whence A is invertible with $A^{-1} = \dfrac{1}{\det A} \text{adj } A$.

Conversely, if A^{-1} exists then, as we have observed above, it follows from Theorem 8.7 that $\det A \neq 0$. □

Theorem 8.9 provides a new way of computing inverses. In purely numerical examples, one can become quite skilful in its use. However, the adjugate matrix has to be constructed with some care! In particular, notice should be taken of the factor $(-1)^{i+j} = \pm 1$. The sign is given according to the scheme

$$+ \ - \ + \ - \ \ldots$$
$$- \ + \ - \ + \ \ldots$$
$$+ \ - \ + \ - \ \ldots$$
$$\vdots \ \ \vdots \ \ \vdots \ \ \vdots$$

Example 8.12

We have seen previously that the matrix

$$A = \begin{bmatrix} 1 & 1 & -1 \\ 2 & 1 & 3 \\ 1 & -5 & 1 \end{bmatrix}$$

is such that $\det A = 28$. By Theorem 8.9, A is therefore invertible. Now the adjugate matrix is

$$\begin{bmatrix} \det \begin{bmatrix} 1 & 3 \\ -5 & 1 \end{bmatrix} & -\det \begin{bmatrix} 1 & -1 \\ -5 & 1 \end{bmatrix} & \det \begin{bmatrix} 1 & -1 \\ 1 & 3 \end{bmatrix} \\ -\det \begin{bmatrix} 2 & 3 \\ 1 & 1 \end{bmatrix} & \det \begin{bmatrix} 1 & -1 \\ 1 & 1 \end{bmatrix} & -\det \begin{bmatrix} 1 & -1 \\ 2 & 3 \end{bmatrix} \\ \det \begin{bmatrix} 2 & 1 \\ 1 & -5 \end{bmatrix} & -\det \begin{bmatrix} 1 & 1 \\ 1 & -5 \end{bmatrix} & \det \begin{bmatrix} 1 & 1 \\ 2 & 1 \end{bmatrix} \end{bmatrix},$$

i.e. it is the following matrix (which with practice can be worked out mentally):

$$B = \begin{bmatrix} 16 & 4 & 4 \\ 1 & 2 & -5 \\ -11 & 6 & -1 \end{bmatrix}.$$

It follows by Theorem 8.9 that $A^{-1} = \dfrac{1}{28} B$, which can of course be verified by direct multiplication.

EXERCISES

8.19 For each of the following matrices, compute its adjugate and then its inverse:

$$\begin{bmatrix} 3 & 1 & 2 \\ 1 & 2 & 1 \\ 1 & 1 & 1 \end{bmatrix}, \quad \begin{bmatrix} 5 & 3 & 2 \\ 2 & 3 & 1 \\ 7 & 5 & 3 \end{bmatrix}, \quad \begin{bmatrix} 1 & 0 & 0 \\ 1 & 2 & 0 \\ 1 & 2 & 3 \end{bmatrix}.$$

8.20 For the matrix
$$A = \begin{bmatrix} -4 & -3 & -3 \\ 1 & 0 & 1 \\ 4 & 4 & 3 \end{bmatrix}$$
show that adj $A = A$.

8.21 If A is an invertible $n \times n$ matrix prove that
$$\det \text{adj } A = (\det A)^{n-1}.$$

8.22 If A and B are invertible $n \times n$ matrices prove that
$$\text{adj } AB = \text{adj } B \cdot \text{adj } A.$$

8.23 If A is an invertible $n \times n$ matrix prove that
$$\text{adj}(\text{adj } A) = (\det A)^{n-2}A.$$
Deduce that, for $n = 2$, $\text{adj}(\text{adj } A) = A$.

8.24 If A is an upper triangular matrix prove that so also is adj A.

8.25 If A is a symmetric matrix prove that so also is adj A.

8.26 If A is an hermitian matrix prove that so also is adj A.

There are other methods of evaluating determinants that are often useful, depending on the matrices involved. For example, there is the so-called 'inspection method' which is best illustrated by example.

Example 8.13

Consider the matrix
$$A = \begin{bmatrix} 1 & x & x^2 \\ 1 & y & y^2 \\ 1 & z & z^2 \end{bmatrix}.$$

Observe that if we set $x = y$ then the first two rows are equal and so the determinant of A reduces to zero. Thus $x - y$ is a factor of det A. Similarly, so are $x - z$ and $y - z$. Consider now the $\sum_{\sigma}$-expansion of det A as in Theorem 8.5. Every term in this expansion consists of a product of entries that come from *distinct* rows and columns (since we are dealing with a permutation σ). Now the highest power of x, y, z appearing in this expansion is 2. Consequently, we can say that
$$\det A = k(x - y)(y - z)(x - z)$$
for some constant k. To determine k, observe that the product of the diagonal entries, namely yz^2, is a term in the $\sum_{\sigma}$-expansion (namely, that which corresponds to the identity permutation). But the term involving yz^2 in the above expression for det A is $-kyz^2$. We conclude therefore that $k = -1$ and so
$$\det A = (x - y)(y - z)(z - x).$$

Example 8.14

Consider the matrix

$$A = \begin{bmatrix} a & b & c & d \\ a^2 & b^2 & c^2 & d^2 \\ b+c+d & c+d+a & d+a+b & a+b+c \\ bcd & cda & dab & abc \end{bmatrix}.$$

By the 'inspection method', factors of det A are

$$a-b, \; a-c, \; a-d, \; b-c, \; b-d, \; c-d.$$

Now the product of the diagonal entries, namely $a^2 b^3 c(d+a+b)$, is a term in the $\sum\limits_{\sigma}$-expansion. But this is only partially represented in the product

$$(a - \underline{b})(\underline{a} - c)(\underline{a} - d)(\underline{b} - c)(\underline{b} - d)(\underline{c} - d),$$

which suggests that we have to find another factor. This can be discovered by adding row 1 to row 3: this clearly produces the factor $a + b + c + d$. Thus we have

$$\det A = k(a-b)(a-c)(a-d)(b-c)(b-d)(c-d)(a+b+c+d)$$

for some constant k. Comparing this with the product $a^2 b^3 c(d+a+b)$ of the diagonal elements, we see that $k = -1$.

Example 8.15

Consider the matrix

$$A = \begin{bmatrix} x & 1 & a & b \\ y^2 & y & 1 & c \\ yz^2 & z^2 & z & 1 \\ yzt & zt & t & 1 \end{bmatrix}.$$

If $x = y$ then the first column is y times the second column whence the determinant is zero and so $x - y$ is a factor of det A. If $z = t$ then the third and fourth rows are the same, so $z - t$ is also a factor of det A. If now $y = z$ then we have

$$\begin{aligned}
\det A &= \det \begin{bmatrix} x & 1 & a & b \\ z^2 & z & 1 & c \\ z^3 & z^2 & z & 1 \\ z^2 t & zt & t & 1 \end{bmatrix} \\[2mm]
&= \det \begin{bmatrix} x & 1 & a & b \\ z^2 & z & 1 & c \\ z^3 & z^2 & z & 1 \\ 0 & 0 & 0 & 1-tc \end{bmatrix} \\[2mm]
&= (1 - tc) \det \begin{bmatrix} x & 1 & a \\ z^2 & z & 1 \\ z^3 & z^2 & z \end{bmatrix} \\[2mm]
&= 0 \qquad \text{since } \rho_3 = z\rho_2.
\end{aligned}$$

Thus we see that $y - z$ is also a factor. It now follows that

$$\det A = k(x - y)(y - z)(z - t)$$

for some constant k, and comparison with the product of the diagonal elements gives $k = 1$.

EXERCISES

8.27 For the matrix

$$A = \begin{bmatrix} 1 & 1 & 1 & 1 \\ a & x & b & c \\ a^2 & x^2 & b^2 & c^2 \\ a^3 & x^3 & b^3 & c^3 \end{bmatrix}$$

express $\det A$ as a product of linear factors.

8.28 Solve the equation

$$\det \begin{bmatrix} x & a & a & a \\ a & x & a & a \\ a & a & x & a \\ a & a & a & x \end{bmatrix} = 0.$$

8.29 Consider the real matrix

$$\begin{bmatrix} x & a & ? & ? \\ y^2 & y & a & ? \\ yz^2 & z^2 & z & a \\ yzt^2 & zt^2 & t^2 & t \end{bmatrix}.$$

Show that, whatever the entries marked ? may be, this matrix has determinant

$$(x - ay)(y - az)(z - at)t.$$

SUPPLEMENTARY EXERCISES

8.30 Let A_n be the $n \times n$ matrix given by

$$a_{ij} = \begin{cases} 0 & \text{if } i = j; \\ 1 & \text{otherwise.} \end{cases}$$

Prove that $\det A_n = (-1)^{n-1}(n - 1)$.

8.31 Consider the $n \times n$ matrix

$$A = \begin{bmatrix} a + b & a & a & \cdots & a \\ a & a + b & a & \cdots & a \\ a & a & a + b & \cdots & a \\ \vdots & \vdots & \vdots & \ddots & \vdots \\ a & a & a & \cdots & a + b \end{bmatrix}.$$

Prove that $\det A = b^{n-1}(na + b)$.

8.32 Solve the equation

$$\det \begin{bmatrix} 1 & 1 & 1 & \ldots & 1 \\ 1 & 1-x & 1 & \ldots & 1 \\ 1 & 1 & 2-x & \ldots & 1 \\ \vdots & \vdots & \vdots & \ddots & \vdots \\ 1 & 1 & 1 & \ldots & n-x \end{bmatrix} = 0.$$

8.33 Consider the $n \times n$ matrix

$$B_n = \begin{bmatrix} b & b & b & \ldots & b & b \\ a & b & b & \ldots & b & b \\ -b & a & b & \ldots & b & b \\ \vdots & \vdots & \vdots & \ddots & \vdots & \vdots \\ -b & -b & -b & \ldots & b & b \\ -b & -b & -b & \ldots & a & b \end{bmatrix}.$$

Prove that

$$\det B_n = (-1)^{n+1}b(a-b)^{n-1}.$$

Hence show that if A_n is the $n \times n$ matrix

$$\begin{bmatrix} a & b & b & \ldots & b \\ -b & a & b & \ldots & b \\ -b & -b & a & \ldots & b \\ \vdots & \vdots & \vdots & \ddots & \vdots \\ -b & -b & -b & \ldots & a \end{bmatrix}.$$

then

$$\det A_n = (a+b)\det A_{n-1} - b(a-b)^{n-1}.$$

Deduce that

$$\det A_n = \tfrac{1}{2}[(a+b)^n + (a-b)^n].$$

8.34 Let A_n be the $n \times n$ matrix given by

$$a_{ij} = \begin{cases} 0 & \text{if } |i-j| > 1; \\ 1 & \text{if } |i-j| = 1; \\ 2\cos\vartheta & \text{if } i = j. \end{cases}$$

If $\Delta_n = \det A_n$, prove that

$$\Delta_{n+2} - 2\cos\vartheta \, \Delta_{n+1} + \Delta_n = 0.$$

Hence show by induction that, for $0 < \vartheta < \pi$,

$$\det A_n = \frac{\sin(n+1)\vartheta}{\sin\vartheta}.$$

8.35 Let A_n be the $n \times n$ matrix given by

$$a_{ij} = \begin{cases} b_i & \text{if } i \neq j; \\ a_i + b_i & \text{if } i = j < n; \\ b_n & \text{if } i = j = n. \end{cases}$$

Prove that $\det A_n = b_n \prod_{i=1}^{n-1} a_i$.

If B_n is given by

$$b_{ij} = \begin{cases} b_i & \text{if } i \neq j; \\ a_i + b_i & \text{if } i = j, \end{cases}$$

prove that $\det B_n = \det A_n + a_n \det B_{n-1}$. Hence show that

$$\det B_n = \prod_{i=1}^{n} a_i + \sum_{i=1}^{n} \left(b_i \prod_{j \neq i} a_j \right).$$

8.36 If A and B are square matrices of the same size, prove that

$$\det \begin{bmatrix} A & B \\ B & A \end{bmatrix} = \det (A + B) \det (A - B).$$

8.37 Let $M = \begin{bmatrix} P & Q \\ R & S \end{bmatrix}$ where P, Q, R, S are square matrices of the same size and P is invertible. Find a matrix N of the form $\begin{bmatrix} A & 0 \\ B & C \end{bmatrix}$ such that

$$NM = \begin{bmatrix} I & P^{-1}Q \\ 0 & S - RP^{-1}Q \end{bmatrix}.$$

Hence show that if P and R commute then $\det M = \det (PS - RQ)$; and that if P and Q commute then $\det M = \det (SP - RQ)$.

8.38 **[Pivotal condensation]** Let $A \in \text{Mat}_{n \times n} \mathbb{R}$ and suppose that $a_{pq} \neq 0$. Let B be the $(n-1) \times (n-1)$ matrix constructed from A by defining

$$b_{ij} = \begin{cases} \det \begin{bmatrix} a_{ij} & a_{iq} \\ a_{pj} & \boxed{a_{pq}} \end{bmatrix} & \text{if } 1 \leqslant i \leqslant p-1, \, 1 \leqslant j \leqslant q-1; \\[12pt] \det \begin{bmatrix} a_{iq} & a_{ij} \\ \boxed{a_{pq}} & a_{pj} \end{bmatrix} & \text{if } 1 \leqslant i \leqslant p-1, \, q+1 \leqslant j \leqslant n; \\[12pt] \det \begin{bmatrix} a_{pj} & \boxed{a_{pq}} \\ a_{ij} & a_{iq} \end{bmatrix} & \text{if } p+1 \leqslant i \leqslant n, \, 1 \leqslant j \leqslant q-1; \\[12pt] \det \begin{bmatrix} \boxed{a_{pq}} & a_{pj} \\ a_{iq} & a_{ij} \end{bmatrix} & \text{if } p+1 \leqslant i \leqslant n, \, q+1 \leqslant j \leqslant n. \end{cases}$$

Prove that $\det A = \dfrac{1}{a_{pq}^{n-2}} \det B$.

[*Hint.* Begin by dividing the p-th row by a_{pq} to obtain a matrix X in which $x_{pq} = 1$. Now subtract suitable multiples of the q-th column of X from the other columns of X to make the elements of the p-th row 0 except for $x_{pq} = 1$, thereby obtaining a matrix Y. Observe how the structure of the matrix B arises. Now consider a Laplace expansion of Y via the p-th row.]

- The $(n-1) \times (n-1)$ matrix B is called *the matrix obtained from A by pivotal condensation using a_{pq} as a pivot*. This useful (and little publicised) result provides a simple recursive way of computing the determinants of matrices and is particularly effective when they have integer entries. The size of the matrix reduces at each step and the calculations are simple since they involve 2×2 submatrices, and are made easier if it can be arranged that a 1 is chosen as a pivot.

For example, $\det \begin{bmatrix} 3 & 2 & 3 \\ 2 & 4 & 3 \\ 4 & 3 & \boxed{2} \end{bmatrix} = \frac{1}{2^{3-2}} \det \begin{bmatrix} -6 & -5 \\ -8 & -1 \end{bmatrix} = -17.$

Compute, via pivotal condensation, the determinants of

$$\begin{bmatrix} 1 & 1 & -1 \\ 2 & 1 & 3 \\ 1 & -5 & 1 \end{bmatrix}, \quad \begin{bmatrix} 1 & 2 & 3 \\ 2 & 3 & 1 \\ 3 & 1 & 2 \end{bmatrix}, \quad \begin{bmatrix} 1 & 2 & 3 & \dots & n \\ -1 & 0 & 3 & \dots & n \\ -1 & -2 & 0 & \dots & n \\ \vdots & \vdots & \vdots & \ddots & \vdots \\ -1 & -2 & -3 & \dots & 0 \end{bmatrix}.$$

9.
Eigenvalues and Eigenvectors

Recall that an $n \times n$ matrix B is **similar** to an $n \times n$ matrix A if there is an invertible $n \times n$ matrix P such that $B = P^{-1}AP$. Our objective now is to determine under what conditions an $n \times n$ matrix is similar to a diagonal matrix. In so doing we shall draw together all of the notions that have been previously developed. Unless otherwise specified, A will denote an $n \times n$ matrix over $\mathbb{R}$ or $\mathbb{C}$.

Definition

By an **eigenvalue** (or **latent root**) of A we shall mean a scalar λ for which there exists a *non-zero* $n \times 1$ matrix $\mathbf{x}$ such that $A\mathbf{x} = \lambda\mathbf{x}$. Such a (column) matrix $\mathbf{x}$ is called an **eigenvector** (or **latent vector**) associated with λ.

- Note that eigenvectors are by definition *non-zero*.

Theorem 9.1

A scalar λ is an eigenvalue of A if and only if
$$\det(A - \lambda I_n) = 0.$$

Proof

Observe that $A\mathbf{x} = \lambda\mathbf{x}$ can be written in the form
$$(A - \lambda I_n)\mathbf{x} = \mathbf{0}.$$
Then λ is an eigenvalue of A if and only if the homogeneous system of equations
$$(A - \lambda I_n)\mathbf{x} = \mathbf{0}$$
has a *non-zero* solution. By Theorems 3.16 and 4.3, this is the case if and only if the matrix $A - \lambda I_n$ is *not* invertible, and by Theorem 8.9 this is equivalent to $\det(A - \lambda I_n)$ being zero. $\square$

Corollary

Similar matrices have the same eigenvalues.

Proof

It suffices to observe that, by Theorem 8.7,

$$\det(P^{-1}AP - \lambda I_n) = \det[P^{-1}(A - \lambda I_n)P]$$
$$= \det P^{-1} \cdot \det(A - \lambda I_n) \cdot \det P$$
$$= \det(A - \lambda I_n). \quad \square$$

Note that with $A = [a_{ij}]_{n \times n}$ we have

$$\det(A - \lambda I_n) = \det \begin{bmatrix} a_{11} - \lambda & a_{12} & \cdots & a_{1n} \\ a_{21} & a_{22} - \lambda & \cdots & a_{2n} \\ \vdots & \vdots & \ddots & \vdots \\ a_{n1} & a_{n2} & \cdots & a_{nn} - \lambda \end{bmatrix}$$

and, recalling that the product of the diagonal elements is a term in the $\sum_{\sigma}$-expansion, we see that this is a polynomial of degree n in λ. We call this the **characteristic polynomial** of A. By the **characteristic equation** of A we mean the equation

$$\det(A - \lambda I_n) = 0.$$

Thus Theorem 9.1 can be expressed by saying that the eigenvalues of A are the roots of the characteristic equation.

Recall that over the field $\mathbb{C}$ of complex numbers this equation has n roots, some of which may be repeated.

If $\lambda_1, \ldots, \lambda_k$ are the distinct roots (= eigenvalues) then the characteristic polynomial factorises in the form

$$(-1)^n (\lambda - \lambda_1)^{r_1} (\lambda - \lambda_2)^{r_2} \cdots (\lambda - \lambda_k)^{r_k}.$$

We call $r_1, \ldots, r_k$ the **algebraic multiplicities** of $\lambda_1, \ldots, \lambda_k$.

Example 9.1

Consider the matrix

$$A = \begin{bmatrix} 0 & 1 \\ -1 & 0 \end{bmatrix}.$$

We have

$$\det(A - \lambda I_2) = \det \begin{bmatrix} -\lambda & 1 \\ -1 & -\lambda \end{bmatrix} = \lambda^2 + 1.$$

Since $\lambda^2 + 1$ has no real roots, we see that A has no real eigenvalues. However, if we regard A as a matrix over $\mathbb{C}$ then A has two eigenvalues, namely i and $-i$, each being of algebraic multiplicity 1.

Example 9.2

Consider the matrix

$$A = \begin{bmatrix} -3 & 1 & -1 \\ -7 & 5 & -1 \\ -6 & 6 & -2 \end{bmatrix}.$$

Using the obvious row/column operations, we compute the characteristic polynomial of A as follows:

$$\det(A - \lambda I_3) = \det \begin{bmatrix} -3-\lambda & 1 & -1 \\ -7 & 5-\lambda & -1 \\ -6 & 6 & -2-\lambda \end{bmatrix}$$

$$= \det \begin{bmatrix} -2-\lambda & 1 & -1 \\ -2-\lambda & 5-\lambda & -1 \\ 0 & 6 & -2-\lambda \end{bmatrix}$$

$$= -(2+\lambda)\det \begin{bmatrix} 1 & 1 & -1 \\ 1 & 5-\lambda & -1 \\ 0 & 6 & -2-\lambda \end{bmatrix}$$

$$= -(2+\lambda)\det \begin{bmatrix} 1 & 1 & -1 \\ 0 & 4-\lambda & 0 \\ 0 & 6 & -2-\lambda \end{bmatrix}$$

$$= -(2+\lambda)(4-\lambda)(-2-\lambda)$$

$$= (2+\lambda)^2(4-\lambda).$$

It follows that the eigenvalues are 4 (of algebraic multiplicity 1) and -2 (of algebraic multiplicity 2).

EXERCISES

9.1 For each of the following matrices, determine the eigenvalues and their algebraic multiplicity:

$$\begin{bmatrix} 1 & 0 & -1 \\ 1 & 2 & 1 \\ 2 & 2 & 3 \end{bmatrix}, \quad \begin{bmatrix} 0 & 1 & 0 \\ 0 & 0 & 1 \\ 1 & -3 & 3 \end{bmatrix}, \quad \begin{bmatrix} 2-i & 0 & i \\ 0 & 1+i & 0 \\ i & 0 & 2-i \end{bmatrix}.$$

9.2 If λ is an eigenvalue of an invertible matrix A prove that $\lambda \neq 0$ and that λ^{-1} is an eigenvalue of A^{-1}.

9.3 Prove that if λ is an eigenvalue of A then, for every polynomial $p(X)$, $p(\lambda)$ is an eigenvalue of $p(A)$.

If λ is an eigenvalue of A then the set

$$E_\lambda = \{\mathbf{x} \in \text{Mat}_{n \times 1} F \; ; \; A\mathbf{x} = \lambda\mathbf{x}\}$$

i.e. the set of eigenvectors associated with the eigenvalue λ together with the zero column $\mathbf{0}$, is readily seen to be a subspace of the vector space $\text{Mat}_{n \times 1} F$.

This subspace of eigenvectors is called the **eigenspace** associated with the eigenvalue λ.

The dimension of the eigenspace E_λ is called the **geometric multiplicity** of the eigenvalue λ.

Example 9.3

Consider the matrix A of the previous example. The eigenvalues are 4 and -2. To determine the eigenspace E_4 we must solve the system $(A - 4I_3)\mathbf{x} = \mathbf{0}$, i.e.

$$\begin{bmatrix} -7 & 1 & -1 \\ -7 & 1 & -1 \\ -6 & 6 & -6 \end{bmatrix} \begin{bmatrix} x \\ y \\ z \end{bmatrix} = \begin{bmatrix} 0 \\ 0 \\ 0 \end{bmatrix}.$$

The corresponding system of equations reduces to $x = 0$, $y - z = 0$ and so E_4 is spanned by

$$\mathbf{x} = \begin{bmatrix} 0 \\ y \\ y \end{bmatrix}$$

where $y \neq 0$ since by definition eigenvectors are non-zero. Consequently we see that the eigenspace E_4 is of dimension 1 with basis

$$\left\{ \begin{bmatrix} 0 \\ 1 \\ 1 \end{bmatrix} \right\}.$$

As for the eigenspace E_{-2}, we solve $(A + 2I_3)\mathbf{x} = \mathbf{0}$, i.e.

$$\begin{bmatrix} -1 & 1 & -1 \\ -7 & 7 & -1 \\ -6 & 6 & 0 \end{bmatrix} \begin{bmatrix} x \\ y \\ z \end{bmatrix} = \begin{bmatrix} 0 \\ 0 \\ 0 \end{bmatrix}.$$

The corresponding system of equations reduces to $x = y$, $z = 0$ and so E_{-2} is spanned by

$$\mathbf{x} = \begin{bmatrix} x \\ x \\ 0 \end{bmatrix}$$

where $x \neq 0$. Thus E_{-2} is also of dimension 1 with basis

$$\left\{ \begin{bmatrix} 1 \\ 1 \\ 0 \end{bmatrix} \right\}.$$

EXERCISES

9.4 For each of the following matrices determine the eigenvalues and a basis
of each of the corresponding eigenspaces:

$$\begin{bmatrix} 1 & 0 & 1 \\ 0 & 1 & 0 \\ 1 & 0 & 1 \end{bmatrix}, \quad \begin{bmatrix} -2 & 5 & 7 \\ 1 & 0 & -1 \\ -1 & 1 & 2 \end{bmatrix}.$$

9.5 Show that the matrix

$$\begin{bmatrix} -2 & -3 & -3 \\ -1 & 0 & -1 \\ 5 & 5 & 6 \end{bmatrix}$$

has only two distinct eigenvalues. Determine a basis of each of the corresponding eigenspaces.

The notions of eigenvalue and eigenvector can also be defined for linear mappings.

Definition

If $f : V \to W$ is linear then a scalar λ is said to be an **eigenvalue** of f if there is a
non-zero $x \in V$ such that $f(x) = \lambda x$, such an element x being called an **eigenvector**
associated with λ.

The connection with matrices is as follows. Given an $n \times n$ matrix A, we can
consider the linear mapping

$$f_A : \mathrm{Mat}_{n \times 1} F \to \mathrm{Mat}_{n \times 1} F$$

given by $f_A(\mathbf{x}) = A\mathbf{x}$. It can readily be verified that, relative to the natural ordered
basis of $\mathrm{Mat}_{n \times 1} F$, we have $\mathrm{Mat}\, f_A = A$. Clearly, the matrix A and the linear mapping
f_A have the same eigenvalues.

Example 9.4

Consider the vector space $\mathrm{Diff}(\mathbb{R}, \mathbb{R})$ of all real differentiable functions. The differentiation map $\mathrm{D} : \mathrm{Diff}(\mathbb{R}, \mathbb{R}) \to \mathrm{Map}(\mathbb{R}, \mathbb{R})$ is linear. An eigenvector of D is a
non-zero differentiable function f such that, for some real λ, $\mathrm{D}f = \lambda f$. By the theory
of first-order differential equations we see that the eigenvectors of D are therefore
the functions f given by $f(x) = ke^{\lambda x}$, where $k \neq 0$ since, we recall, eigenvectors are
by definition non-zero.

Example 9.5

Consider the linear mapping $f : \mathbb{R}^3 \to \mathbb{R}^3$ given by

$$f(x, y, z) = (y + z, \ x + z, \ x + y).$$

Relative to the natural ordered basis of $\mathbb{R}^3$ the matrix of f is

$$A = \begin{bmatrix} 0 & 1 & 1 \\ 1 & 0 & 1 \\ 1 & 1 & 0 \end{bmatrix}.$$

The reader can readily verify that

$$\det (A - \lambda I_3) = \det \begin{bmatrix} -\lambda & 1 & 1 \\ 1 & -\lambda & 1 \\ 1 & 1 & -\lambda \end{bmatrix} = -(\lambda + 1)^2 (\lambda - 2).$$

The eigenvalues of A, and hence those of f, are therefore 2 and -1, the latter being of algebraic multiplicity 2.

EXERCISES

9.6 Determine the eigenvalues, and their algebraic multiplicities, of the linear mapping $f : \mathbb{R}^3 \to \mathbb{R}^3$ given by

(1) $f(x, y, z) = (x + 2y + 2z, \ 2y + z, \ -x + 2y + 2z)$;

(2) $f(x, y, z) = (y + z, \ 0, \ x + y)$.

Theorem 9.2

Eigenvectors corresponding to distinct eigenvalues are linearly independent.

Proof

The proof is by induction. If $f : V \to V$ has only one eigenvalue and if x is a corresponding eigenvector then since $x \neq 0$ we know that $\{x\}$ is linearly independent. For the inductive step, suppose that every set of n eigenvectors that correspond to n distinct eigenvalues is linearly independent. Let $x_1, \ldots, x_{n+1}$ be eigenvectors that correspond to distinct eigenvalues $\lambda_1, \ldots, \lambda_{n+1}$. If we have

(1) $$a_1 x_1 + \cdots + a_n x_n + a_{n+1} x_{n+1} = 0$$

then, applying f and using the fact that $f(x_i) = \lambda_i x_i$, we obtain

(2) $$a_1 \lambda_1 x_1 + \cdots + a_n \lambda_n x_n + a_{n+1} \lambda_{n+1} x_{n+1} = 0.$$

Now take $(2) - \lambda_{n+1}(1)$ to get

$$a_1 (\lambda_1 - \lambda_{n+1}) x_1 + \cdots + a_n (\lambda_n - \lambda_{n+1}) x_n = 0.$$

By the induction hypothesis and the fact that $\lambda_1, \ldots, \lambda_{n+1}$ are distinct, we deduce that

$$a_1 = \cdots = a_n = 0.$$

It now follows by (1) that $a_{n+1} x_{n+1} = 0$ whence, since $x_{n+1} \neq 0$, we also have $a_{n+1} = 0$. Hence $x_1, \ldots, x_{n+1}$ are linearly independent and the result follows. $\square$

Definition

A linear mapping $f : V \to V$ is said to be **diagonalisable** if there is an ordered basis $(v_i)_n$ of V with repect to which the matrix of f is a diagonal matrix.

Thus f is diagonalisable if and only if there is an ordered basis $(v_i)_n$ of V such that

$$
\begin{aligned}
f(v_1) &= \lambda_1 v_1 \\
f(v_2) &= \lambda_2 v_2 \\
& \vdots \ddots \\
f(v_n) &= \lambda_n v_n
\end{aligned}
$$

in which case the λ_i are the eigenvalues of f. We can therefore assert the following:

Theorem 9.3

A linear mapping $f : V \to V$ is diagonalisable if and only if V has a basis consisting of eigenvectors of f. $\square$

Equivalently, if we define a square matrix to be **diagonalisable** when it is similar to a diagonal matrix then as a result on matrices Theorem 9.3 translates into the following result:

Theorem 9.4

An $n \times n$ matrix is diagonalisable if and only if it admits n linearly independent eigenvectors. $\square$

We now proceed to show that $f : V \to V$ is diagonalisable if and only if, for every eigenvalue λ, the geometric and algebraic multiplicities of λ coincide. For this purpose we require the following results.

Theorem 9.5

Let V be of dimension n. If $\lambda_1, \ldots, \lambda_k$ are the eigenvalues of $f : V \to V$ and if $d_1, \ldots, d_k$ are their geometric multiplicities then

$$d_1 + \cdots + d_k \leqslant n$$

with equality if and only if f is diagonalisable.

Proof

For each i let B_i be a basis of E_{λ_i}. Observe first that if

$$v_1 + \cdots + v_k = 0$$

where each $v_i \in B_i$ then necessarily each $v_i = 0$. This follows from Theorem 9.2.

Observe next that $\bigcup_{i=1}^{k} B_i$ is linearly independent. In fact, if $B_i = \{e_{i1}, \ldots, e_{id_i}\}$ and

$$v_i = \mu_{i1} e_{i1} + \cdots + \mu_{id_i} e_{id_i} \in B_i$$

then $\sum_{i=1}^{k} v_i = 0$ gives, from the above, each $v_i = 0$ whence all the coefficients $\mu_{ij} = 0$.

Since the B_i are pairwise disjoint, it follows that

$$d_1 + \cdots + d_k = \left| \bigcup_{i=1}^{k} B_i \right| \leqslant n.$$

Finally, equality occurs if and only if V has n linearly independent eigenvectors, i.e. by Theorem 9.3, if and only if f is diagonalisable. $\quad\square$

Theorem 9.6

If λ is an eigenvalue of $f : V \to V$ then the geometric multiplicity of λ is less than or equal to the algebraic multiplicity of λ.

Proof

Let $\{e_1, \ldots, e_d\}$ be a basis of E_λ and extend this to a basis $B = \{e_1, \ldots, e_n\}$ of V. The matrix of f relative to B is of the form

$$M = \begin{bmatrix} \lambda I_d & C \\ 0 & D \end{bmatrix}.$$

The characteristic polynomial of M is of the form $(\lambda - X)^d p(X)$ where $p(X)$ is a polynomial of degree $n - d$. It follows that d is less than or equal to the algebraic multiplicity of λ. $\quad\square$

We can now deduce from the above the following necessary and sufficient condition for $f : V \to V$ to be diagonalisable.

Theorem 9.7

The following statements are equivalent:

(1) $f : V \to V$ is diagonalisable;

(2) for every eigenvalue λ of f, the geometric multiplicity of λ coincides with the algebraic multiplicity of λ.

Proof

The sum of the algebraic multiplicities of the eigenvalues is the degree of the characteristic polynomial, namely $n = \dim V$. The result therefore follows from Theorems 9.5 and 9.6. $\quad\square$

Example 9.6

As observed in Example 9.2 above, the matrix

$$A = \begin{bmatrix} -3 & 1 & -1 \\ -7 & 5 & -1 \\ -6 & 6 & -2 \end{bmatrix}$$

has only two distinct eigenvalues, namely 4 and -2. The latter is of algebraic multiplicity 2. To determine the eigenspace E_{-2}, we solve $(A + 2I_3)\mathbf{x} = \mathbf{0}$, i.e.

$$\begin{bmatrix} -1 & 1 & -1 \\ -7 & 7 & -1 \\ -6 & 6 & 0 \end{bmatrix} \begin{bmatrix} x \\ y \\ z \end{bmatrix} = \begin{bmatrix} 0 \\ 0 \\ 0 \end{bmatrix}.$$

The corresponding system of equations reduces to

$$x - y = 0,$$

$$z = 0$$

so the rank of the coefficient matrix is 2 and consequently the solution space is of dimension $3 - 2 = 1$. Thus the eigenvalue -2 is of geometric multiplicity 1.

It follows by Theorem 9.7 that A is not diagonalisable.

Example 9.7

Consider the matrix

$$B = \begin{bmatrix} 1 & -3 & 3 \\ 3 & -5 & 3 \\ 6 & -6 & 4 \end{bmatrix}.$$

The reader can readily verify that

$$\det (B - \lambda I_3) = (4 - \lambda)(\lambda + 2)^2,$$

so the eigenvalues are 4 and -2, these being of respective algebraic multiplicities 1 and 2.

To determine the eigenspace E_{-2} we solve $(B + 2I_3)\mathbf{x} = \mathbf{0}$, i.e.

$$\begin{bmatrix} 3 & -3 & 3 \\ 3 & -3 & 3 \\ 6 & -6 & 6 \end{bmatrix} \begin{bmatrix} x \\ y \\ z \end{bmatrix} = \begin{bmatrix} 0 \\ 0 \\ 0 \end{bmatrix}.$$

The corresponding system of equations reduces to $x - y + z = 0$, so the coefficient matrix is of rank 1 and so the dimension of the solution space is $3 - 1 = 2$. Thus the eigenvalue -2 is of geometric multiplicity 2.

As for the eigenvalue 4, since its algebraic multiplicity is 1, it follows by Theorem 9.6 that its geometric multiplicity is also 1. It now follows by Theorem 9.7 that B is diagonalisable.

If A is similar to a diagonal matrix D then there is an invertible matrix P such that $P^{-1}AP = D$ where the diagonal entries of D are the eigenvalues of A. We shall now consider the problem of determining such a matrix P.

First we observe that the equation $P^{-1}AP = D$ can be written $AP = PD$. Let the columns of P be $\mathbf{p}_1, \ldots, \mathbf{p}_n$ and let

$$D = \begin{bmatrix} \lambda_1 & & & \\ & \lambda_2 & & \\ & & \ddots & \\ & & & \lambda_n \end{bmatrix}$$

where $\lambda_1, \ldots, \lambda_n$ are the eigenvalues of A. Comparing the i-th columns of each side of the equation $AP = PD$, we obtain

$$(i = 1, \ldots, n) \qquad A\mathbf{p}_i = \lambda_i \mathbf{p}_i.$$

In other words, *the i-th column of P is an eigenvector of A corresponding to the eigenvalue λ_i.*

Example 9.8

Consider again the previous example. Any two linearly independent eigenvectors in E_{-2} constitute a basis for E_{-2}. For these we can choose, for example,

$$\begin{bmatrix} 1 \\ 1 \\ 0 \end{bmatrix}, \quad \begin{bmatrix} 1 \\ 0 \\ -1 \end{bmatrix}.$$

Any single non-zero vector in E_4 constitutes a basis for E_4. We can choose, for example,

$$\begin{bmatrix} 1 \\ 1 \\ 2 \end{bmatrix}.$$

Clearly, the three eigenvectors

$$\begin{bmatrix} 1 \\ 1 \\ 0 \end{bmatrix}, \quad \begin{bmatrix} 1 \\ 0 \\ -1 \end{bmatrix}, \quad \begin{bmatrix} 1 \\ 1 \\ 2 \end{bmatrix}$$

are linearly independent. Pasting these eigenvectors together, we obtain the matrix

$$P = \begin{bmatrix} 1 & 1 & 1 \\ 1 & 0 & 1 \\ 0 & -1 & 2 \end{bmatrix},$$

and this is such that

$$P^{-1}BP = \begin{bmatrix} -2 & 0 & 0 \\ 0 & -2 & 0 \\ 0 & 0 & 4 \end{bmatrix}.$$

- Note that, in order to obtain a particular arrangement of the eigenvalues down the diagonal of the matrix D, it suffices to select the same arrangement of the columns of P.

EXERCISES

9.7 For each of the matrices A given by

$$\begin{bmatrix} 1 & 0 & 1 \\ 0 & 1 & 0 \\ 1 & 0 & 1 \end{bmatrix}, \quad \begin{bmatrix} -2 & 5 & 7 \\ 1 & 0 & -1 \\ -1 & 1 & 2 \end{bmatrix},$$

$$\begin{bmatrix} -3 & -7 & 19 \\ -2 & -1 & 8 \\ -2 & -3 & 10 \end{bmatrix}, \quad \begin{bmatrix} -4 & 0 & -3 \\ 1 & 3 & 1 \\ 4 & -2 & 2 \end{bmatrix}$$

find a matrix P such that $P^{-1}AP$ is diagonal.

Let us now return to the problems of equilibrium-seeking systems and difference equations as outlined in Chapter 2. In each of these, the matrix in question is of size 2×2, so we first prove a simple result that will allow us to cut a few corners.

Theorem 9.8

If the 2×2 matrix

$$A = \begin{bmatrix} a & b \\ c & d \end{bmatrix}$$

has distinct eigenvalues λ_1, λ_2 then it is diagonalisable. When $b \neq 0$, an invertible matrix P such that

$$P^{-1}AP = \begin{bmatrix} \lambda_1 & 0 \\ 0 & \lambda_2 \end{bmatrix}$$

is the matrix

$$P = \begin{bmatrix} b & b \\ \lambda_1 - a & \lambda_2 - a \end{bmatrix}.$$

Proof

The first statement is immediate from Theorems 9.2 and 9.4. As for the second statement, we observe that

$$\det \begin{bmatrix} a - \lambda & b \\ c & d - \lambda \end{bmatrix} = \lambda^2 - (a + d)\lambda + ad - bc$$

and so the eigenvalues of A are

$$\lambda_1 = \tfrac{1}{2}[(a + d) + \sqrt{(a - d)^2 + 4bc}],$$
$$\lambda_2 = \tfrac{1}{2}[(a + d) - \sqrt{(a - d)^2 + 4bc}].$$

Consider the column matrix

$$\mathbf{x}_1 = \begin{bmatrix} b \\ \lambda_1 - a \end{bmatrix}$$

in which, by hypothesis, $b \neq 0$. We have

$$\begin{bmatrix} a & b \\ c & d \end{bmatrix} \begin{bmatrix} b \\ \lambda_1 - a \end{bmatrix} = \begin{bmatrix} b\lambda_1 \\ cb + d(\lambda_1 - a) \end{bmatrix} = \lambda_1 \begin{bmatrix} b \\ \lambda_1 - a \end{bmatrix},$$

the final equality resulting from the fact that

$$\lambda_1(\lambda_1 - a) - cb - d(\lambda_1 - a) = \lambda_1^2 - (a + d)\lambda_1 + ad - bc = 0.$$

Thus $\mathbf{x}_1$ is an eigenvector associated with λ_1. Similarly, we can show that

$$\mathbf{x}_2 = \begin{bmatrix} b \\ \lambda_2 - a \end{bmatrix}$$

is an eigenvector associated with λ_2. Pasting these eigenvectors together, we obtained the required matrix P. $\square$

Example 9.9

Consider the equilibrium-seeking system as described in Chapter 2. The matrix in question is

$$\begin{bmatrix} \frac{1}{4} & \frac{1}{20} \\ \frac{3}{4} & \frac{19}{20} \end{bmatrix}.$$

The eigenvalues of A are the roots of the equation

$$(\tfrac{1}{4} - \lambda)(\tfrac{19}{20} - \lambda) - \tfrac{3}{80} = 0.$$

The reader will easily check that this reduces to

$$(5\lambda - 1)(\lambda - 1) = 0$$

so that the eigenvalues are $\frac{1}{5}$ and 1. It follows by Theorem 9.8 that A is diagonalisable, that an eigenvector associated with $\lambda_1 = \frac{1}{5}$ is

$$\begin{bmatrix} \frac{1}{20} \\ \frac{1}{5} - \frac{1}{4} \end{bmatrix} = \begin{bmatrix} \frac{1}{20} \\ -\frac{1}{20} \end{bmatrix},$$

and that an eigenvector associated with $\lambda_2 = 1$ is

$$\begin{bmatrix} \frac{1}{20} \\ 1 - \frac{1}{4} \end{bmatrix} = \begin{bmatrix} \frac{1}{20} \\ \frac{3}{4} \end{bmatrix}.$$

We can therefore assert that the matrix $P = \begin{bmatrix} 1 & 1 \\ -1 & 15 \end{bmatrix}$ is invertible and such that

$$P^{-1}AP = \begin{bmatrix} \frac{1}{5} & 0 \\ 0 & 1 \end{bmatrix}.$$

Since, as is readily seen,

$$P^{-1} = \tfrac{1}{16} \begin{bmatrix} 15 & -1 \\ 1 & 1 \end{bmatrix},$$

we can compute $A^n = P \begin{bmatrix} \tfrac{1}{5} & 0 \\ 0 & 1 \end{bmatrix}^n P^{-1}$. We have

$$\begin{aligned} A^n &= \tfrac{1}{16} \begin{bmatrix} 1 & 1 \\ -1 & 15 \end{bmatrix} \begin{bmatrix} \tfrac{1}{5^n} & 0 \\ 0 & 1 \end{bmatrix} \begin{bmatrix} 15 & -1 \\ 1 & 1 \end{bmatrix} \\ &= \tfrac{1}{16} \begin{bmatrix} 1 & 1 \\ -1 & 15 \end{bmatrix} \begin{bmatrix} \tfrac{15}{5^n} & -\tfrac{1}{5^n} \\ 1 & 1 \end{bmatrix} \\ &= \tfrac{1}{16} \begin{bmatrix} 1 + \tfrac{15}{5^n} & 1 - \tfrac{1}{5^n} \\ 15(1 - \tfrac{1}{5^n}) & 15 + \tfrac{1}{5^n} \end{bmatrix}. \end{aligned}$$

Example 9.10

Consider the **Fibonacci sequence** $(a_i)_{i \geqslant 0}$ defined recursively by $a_0 = 0$, $a_1 = 1$, and

$$(\forall n \geqslant 0) \quad a_{n+2} = a_{n+1} + a_n.$$

We can write this as a system of difference equations in the following way:

$$a_{n+2} = a_{n+1} + b_{n+1}$$
$$b_{n+2} = a_{n+1}.$$

This we can represent in the matrix form $\mathbf{x}_{n+2} = A\mathbf{x}_{n+1}$ where

$$\mathbf{x}_n = \begin{bmatrix} a_n \\ b_n \end{bmatrix} \quad \text{and} \quad A = \begin{bmatrix} 1 & 1 \\ 1 & 0 \end{bmatrix}.$$

The eigenvalues of A are the solutions of $\lambda^2 - \lambda - 1 = 0$, namely

$$\lambda_1 = \tfrac{1}{2}(1 + \sqrt{5}), \quad \lambda_2 = \tfrac{1}{2}(1 - \sqrt{5}).$$

By Theorem 9.8, A is diagonalisable, and corresponding eigenvectors are

$$\begin{bmatrix} 1 \\ \lambda_1 - 1 \end{bmatrix} = \begin{bmatrix} 1 \\ -\lambda_2 \end{bmatrix}, \qquad \begin{bmatrix} 1 \\ \lambda_2 - 1 \end{bmatrix} = \begin{bmatrix} 1 \\ -\lambda_1 \end{bmatrix}.$$

Then the matrix $P = \begin{bmatrix} 1 & 1 \\ -\lambda_2 & -\lambda_1 \end{bmatrix}$ is invertible and such that

$$P^{-1}AP = \begin{bmatrix} \lambda_1 & 0 \\ 0 & \lambda_2 \end{bmatrix}.$$

Now clearly

$$P^{-1} = \tfrac{1}{\lambda_2 - \lambda_1} \begin{bmatrix} -\lambda_1 & -1 \\ \lambda_2 & 1 \end{bmatrix}$$

and so, using the fact that $\lambda_1 \lambda_2 = -1$, we can compute

$$A^n = \frac{1}{\lambda_2 - \lambda_1} \begin{bmatrix} 1 & 1 \\ -\lambda_2 & -\lambda_1 \end{bmatrix} \begin{bmatrix} \lambda_1^n & 0 \\ 0 & \lambda_2^n \end{bmatrix} \begin{bmatrix} -\lambda_1 & -1 \\ \lambda_2 & 1 \end{bmatrix}$$

$$= \frac{1}{\lambda_2 - \lambda_1} \begin{bmatrix} 1 & 1 \\ -\lambda_2 & -\lambda_1 \end{bmatrix} \begin{bmatrix} -\lambda_1^{n+1} & -\lambda_1^n \\ \lambda_2^{n+1} & \lambda_2^n \end{bmatrix}$$

$$= \frac{1}{\lambda_2 - \lambda_1} \begin{bmatrix} \lambda_2^{n+1} - \lambda_1^{n+1} & \lambda_2^n - \lambda_1^n \\ \lambda_2^n - \lambda_1^n & \lambda_2^{n-1} - \lambda_1^{n-1} \end{bmatrix}.$$

Since it is given that $b_1 = a_0 = 0$ and $a_1 = 1$, we can now assert that

$$\begin{bmatrix} a_{n+1} \\ b_{n+1} \end{bmatrix} = A^n \mathbf{x}_1 = A^n \begin{bmatrix} 1 \\ 0 \end{bmatrix} = \frac{1}{\lambda_2 - \lambda_1} \begin{bmatrix} \lambda_2^{n+1} - \lambda_1^{n+1} \\ \lambda_2^n - \lambda_1^n \end{bmatrix}$$

and hence we see that

$$a_n = \frac{1}{\lambda_2 - \lambda_1}(\lambda_2^n - \lambda_1^n) = \frac{1}{\sqrt{5}}[\tfrac{1}{2}(1 + \sqrt{5})]^n - \frac{1}{\sqrt{5}}[\tfrac{1}{2}(1 - \sqrt{5})]^n.$$

Example 9.11

Consider the sequence of fractions

$$2, \quad 2 + \tfrac{1}{2}, \quad 2 + \frac{1}{2 + \tfrac{1}{2}}, \quad 2 + \cfrac{1}{2 + \cfrac{1}{2 + \tfrac{1}{2}}}, \quad \cdots,$$

This is a particular example of what is called a **continued fraction**. If we denote the n-th term in this sqeuence by $\dfrac{a_n}{b_n}$ then we have

$$\frac{a_{n+1}}{b_{n+1}} = 2 + \frac{1}{\dfrac{a_n}{b_n}} = \frac{2a_n + b_n}{a_n}$$

and so we can consider the difference equations

$$a_{n+1} = 2a_n + b_n$$
$$b_{n+1} = a_n.$$

The matrix of the system is

$$A = \begin{bmatrix} 2 & 1 \\ 1 & 0 \end{bmatrix}$$

and we can write the system as $\mathbf{x}_{n+1} = A\mathbf{x}_n$ where

$$\mathbf{x}_n = \begin{bmatrix} a_n \\ b_n \end{bmatrix}.$$

Now $a_1 = 2$ and $b_1 = 1$, so we can compute $\mathbf{x}_{n+1}$ from

$$\mathbf{x}_{n+1} = A^n \begin{bmatrix} 2 \\ 1 \end{bmatrix}.$$

The eigenvalues of A are the solutions of $\lambda^2 - 2\lambda - 1 = 0$, namely
$$\lambda_1 = 1 + \sqrt{2}, \quad \lambda_2 = 1 - \sqrt{2}.$$
By Theorem 9.8, the matrix
$$P = \begin{bmatrix} 1 & 1 \\ -1 + \sqrt{2} & -1 - \sqrt{2} \end{bmatrix} = \begin{bmatrix} 1 & 1 \\ -\lambda_2 & -\lambda_1 \end{bmatrix}$$
is invertible and such that
$$P^{-1}AP = \begin{bmatrix} 1 + \sqrt{2} & 0 \\ 0 & 1 - \sqrt{2} \end{bmatrix}.$$
Now it is readily seen that
$$P^{-1} = \tfrac{1}{2\sqrt{2}} \begin{bmatrix} 1 + \sqrt{2} & 1 \\ -1 + \sqrt{2} & -1 \end{bmatrix} = \tfrac{1}{2\sqrt{2}} \begin{bmatrix} \lambda_1 & 1 \\ -\lambda_2 & -1 \end{bmatrix}.$$
Consequently,
$$A^n = \tfrac{1}{2\sqrt{2}} \begin{bmatrix} 1 & 1 \\ -\lambda_2 & -\lambda_1 \end{bmatrix} \begin{bmatrix} \lambda_1^n & 0 \\ 0 & \lambda_2^n \end{bmatrix} \begin{bmatrix} \lambda_1 & 1 \\ -\lambda_2 & -1 \end{bmatrix}$$
$$= \tfrac{1}{2\sqrt{2}} \begin{bmatrix} 1 & 1 \\ -\lambda_2 & -\lambda_1 \end{bmatrix} \begin{bmatrix} \lambda_1^{n+1} & \lambda_1^n \\ -\lambda_2^{n+1} & -\lambda_2^n \end{bmatrix}$$
$$= \tfrac{1}{2\sqrt{2}} \begin{bmatrix} \lambda_1^{n+1} - \lambda_2^{n+1} & \lambda_1^n - \lambda_2^n \\ \lambda_1^n - \lambda_2^n & \lambda_1^{n-1} - \lambda_2^{n-1} \end{bmatrix}$$
and so we deduce from
$$\begin{bmatrix} a_{n+1} \\ b_{n+1} \end{bmatrix} = A^n \begin{bmatrix} 2 \\ 1 \end{bmatrix}$$
that the general term $\dfrac{a_{n+1}}{b_{n+1}}$ is given by
$$\frac{2[(1 + \sqrt{2})^{n+1} - (1 - \sqrt{2})^{n+1}] + (1 + \sqrt{2})^n - (1 - \sqrt{2})^n}{2[(1 + \sqrt{2})^n - (1 - \sqrt{2})^n] + (1 + \sqrt{2})^{n-1} - (1 - \sqrt{2})^{n-1}}.$$

Example 9.12

If q_1 is a positive rational, define
$$q_2 = \frac{2 + q_1}{1 + q_1}.$$
Then it is easy to show that
$$|2 - q_2^2| < |2 - q_1^2|.$$
In other words, if q_1 is an approximation to $\sqrt{2}$ then q_2 is a better approximation. Starting with $q_1 = 1$ and applying this observation repeatedly, we obtain the sequence
$$1, \quad \frac{3}{2}, \quad \frac{7}{5}, \quad \frac{17}{12}, \quad \frac{41}{29}, \quad \frac{99}{70}, \quad \ldots .$$

We can use the techniques described above to determine the general term in this sequence and show that it does indeed converge to $\sqrt{2}$. Denoting the n-th term by $\dfrac{a_n}{b_n}$, we have

$$\frac{a_{n+1}}{b_{n+1}} = \frac{2 + \dfrac{a_n}{b_n}}{1 + \dfrac{a_n}{b_n}} = \frac{2b_n + a_n}{b_n + a_n}$$

and so the sequence can be described by the system of difference equations

$$a_{n+1} = a_n + 2b_n$$
$$b_{n+1} = a_n + b_n.$$

The matrix of the system is

$$A = \begin{bmatrix} 1 & 2 \\ 1 & 1 \end{bmatrix}$$

and its characteristic equation is $\lambda^2 - 2\lambda - 1 = 0$, so that the eigenvalues are

$$\lambda_1 = 1 + \sqrt{2}, \quad \lambda_2 = 1 - \sqrt{2}.$$

By Theorem 9.8, the matrix

$$P = \begin{bmatrix} 2 & 2 \\ \sqrt{2} & -\sqrt{2} \end{bmatrix}$$

is invertible and such that

$$P^{-1}AP = \begin{bmatrix} 1 + \sqrt{2} & 0 \\ 0 & 1 - \sqrt{2} \end{bmatrix}.$$

Now it is readily seen that

$$P^{-1} = \tfrac{1}{4}\begin{bmatrix} 1 & \sqrt{2} \\ 1 & -\sqrt{2} \end{bmatrix}.$$

Consequently,

$$A^n = \tfrac{1}{4}\begin{bmatrix} 2 & 2 \\ \sqrt{2} & -\sqrt{2} \end{bmatrix}\begin{bmatrix} \lambda_1^n & 0 \\ 0 & \lambda_2^n \end{bmatrix}\begin{bmatrix} 1 & \sqrt{2} \\ 1 & -\sqrt{2} \end{bmatrix}$$

$$= \tfrac{1}{4}\begin{bmatrix} 2 & 2 \\ \sqrt{2} & -\sqrt{2} \end{bmatrix}\begin{bmatrix} \lambda_1^n & \sqrt{2}\lambda_1^n \\ \lambda_2^n & -\sqrt{2}\lambda_2^n \end{bmatrix}$$

$$= \tfrac{1}{4}\begin{bmatrix} 2\lambda_1^n + 2\lambda_2^n & 2\sqrt{2}\lambda_1^n - 2\sqrt{2}\lambda_2^n \\ \sqrt{2}\lambda_1^n - \sqrt{2}\lambda_2^n & 2\lambda_1^n + 2\lambda_2^n \end{bmatrix}$$

and so we deduce from

$$\begin{bmatrix} a_{n+1} \\ b_{n+1} \end{bmatrix} = A^n \begin{bmatrix} 1 \\ 1 \end{bmatrix}$$

that

$$\frac{a_{n+1}}{b_{n+1}} = \frac{2(1 + \sqrt{2})^{n+1} + 2(1 - \sqrt{2})^{n+1}}{\sqrt{2}(1 + \sqrt{2})^{n+1} - \sqrt{2}(1 - \sqrt{2})^{n+1}} = \sqrt{2} \cdot \frac{1 + \left(\frac{1-\sqrt{2}}{1+\sqrt{2}}\right)^{n+1}}{1 - \left(\frac{1-\sqrt{2}}{1+\sqrt{2}}\right)^{n+1}},$$

from which we see that

$$\lim_{n\to\infty} \frac{a_n}{b_n} = \sqrt{2}.$$

It is of course possible for problems such as the above to involve a (non-diagonal) 2×2 matrix A whose eigenvalues are not distinct. In this case A is not diagonalisable; for if λ is the only eigenvalue then the system of equations $(A - \lambda I_2)\mathbf{x} = \mathbf{0}$ reduces to a single equation and the dimension of the solution space is $2 - 1 = 1$, so there cannot exist two linearly independent eigenvectors. To find high powers of A in this case we have to proceed in a different manner. If

$$A = \begin{bmatrix} a & b \\ c & d \end{bmatrix}$$

then the characteristic polynomial of A is

$$f(X) = X^2 - (a + d)X + ad - bc.$$

Observe now that

$$A^2 = \begin{bmatrix} a^2 + bc & b(a + d) \\ c(a + d) & bc + d^2 \end{bmatrix}$$

$$= (a + d)\begin{bmatrix} a & b \\ c & d \end{bmatrix} - (ad - bc)\begin{bmatrix} 1 & 0 \\ 0 & 1 \end{bmatrix}$$

$$= (a + d)A - (ad - bc)I_2$$

and so we see that $f(A) = 0$. For $n \geqslant 2$ consider the euclidean division of X^n by $f(X)$. Since f is of degree 2 we have

(3) $$X^n = f(X)q(X) + \alpha_1 X + \alpha_2.$$

Substituting A for X in this polynomial identity we obtain, by the above observation,

$$A^n = \alpha_1 A + \alpha_2 I_2.$$

We can determine α_1 and α_2 as follows. If we differentiate (3) and substitute λ (the single eigenvalue of A) for X then, since $f(\lambda) = 0$, we obtain

$$n\lambda^{n-1} = \alpha_1.$$

Also, substituting λ for X in (3) and again using $f(\lambda) = 0$, we obtain

$$\lambda^n = \alpha_1\lambda + \alpha_2 = n\lambda^n + \alpha_2$$

and so

$$\alpha_2 = (1 - n)\lambda^n.$$

It now follows that

$$A^n = n\lambda^{n-1}A + (1 - n)\lambda^n I_2.$$

Example 9.13

Consider the $n \times n$ **tridiagonal** matrix

$$
A_n = \begin{bmatrix}
2 & 1 & 0 & 0 & \cdots & 0 & 0 \\
1 & 2 & 1 & 0 & \cdots & 0 & 0 \\
0 & 1 & 2 & 1 & \cdots & 0 & 0 \\
\vdots & \vdots & \vdots & \vdots & \ddots & \vdots & \vdots \\
0 & 0 & 0 & 0 & \cdots & 2 & 1 \\
0 & 0 & 0 & 0 & \cdots & 1 & 2
\end{bmatrix}.
$$

Writing $a_n = \det A_n$ we have, using a Laplace expansion along the first row,

$$
a_n = 2a_{n-1} - \det \begin{bmatrix}
1 & 1 & 0 & 0 & \cdots & 0 \\
0 & 2 & 1 & 0 & \cdots & 0 \\
0 & 1 & 2 & 1 & \cdots & 0 \\
\vdots & \vdots & \vdots & \vdots & \ddots & \vdots \\
0 & 0 & 0 & 0 & \cdots & 2
\end{bmatrix}
$$

$$
= 2a_{n-1} - a_{n-2}.
$$

Expressing this recurrence relation in the usual way as a system of difference equations

$$
a_n = 2a_{n-1} - b_{n-1}
$$

$$
b_n = a_{n-1}
$$

we consider the system $\mathbf{x}_n = A\mathbf{x}_{n-1}$ where

$$
\mathbf{x} = \begin{bmatrix} a_n \\ b_n \end{bmatrix} \quad \text{and} \quad A = \begin{bmatrix} 2 & -1 \\ 1 & 0 \end{bmatrix}.
$$

Now

$$
\det (A - \lambda I_2) = \lambda(\lambda - 2) + 1 = (\lambda - 1)^2,
$$

and so A has the single eigenvalue 1 of algebraic multiplicity 2. We can compute A^n as in the above:

$$
A^n = nA + (1 - n)I_2 = \begin{bmatrix} n + 1 & -n \\ n & 1 - n \end{bmatrix}.
$$

Consequently we have

$$
\begin{bmatrix} a_n \\ b_n \end{bmatrix} = A^{n-2} \begin{bmatrix} a_2 \\ b_2 \end{bmatrix}
$$

$$
= \begin{bmatrix} n - 1 & -n + 2 \\ n - 2 & 3 - n \end{bmatrix} \begin{bmatrix} 3 \\ 2 \end{bmatrix}
$$

$$
= \begin{bmatrix} n + 1 \\ n \end{bmatrix}
$$

and hence we see that

$$\det A_n = a_n = n + 1.$$

EXERCISES

9.8 Consider the tridiagonal matrix

$$A_n = \begin{bmatrix} 1 & -4 & 0 & \dots & 0 & 0 \\ 5 & 1 & -4 & \dots & 0 & 0 \\ 0 & 5 & 1 & \dots & 0 & 0 \\ \vdots & \vdots & \vdots & \ddots & \vdots & \vdots \\ 0 & 0 & 0 & \dots & 1 & -4 \\ 0 & 0 & 0 & \dots & 5 & 1 \end{bmatrix}.$$

Prove that

$$\det A_n = \tfrac{1}{9}[5^{n+1} - (-4)^{n+1}].$$

SUPPLEMENTARY EXERCISES

9.9 Determine the characteristic polynomials of

$$\begin{bmatrix} 1 & 2 & 3 \\ 0 & 1 & 2 \\ 0 & 0 & 1 \end{bmatrix}, \quad \begin{bmatrix} 1 & 1 & 0 \\ -1 & 1 & 1 \\ 0 & 1 & -1 \end{bmatrix}, \quad \begin{bmatrix} 1 & -4 & 0 \\ 2 & -2 & -2 \\ -2\tfrac{1}{2} & 1 & -2 \end{bmatrix}.$$

9.10 Suppose that A and B are $n \times n$ matrices such that $I_n - AB$ is invertible. Prove that so also is $I_n - BA$ with

$$(I_n - BA)^{-1} = I_n + B(I_n - AB)^{-1}A.$$

Deduce that XY and YX have the same eigenvalues.

9.11 Show that $\begin{bmatrix} 1 \\ i \end{bmatrix}$ and $\begin{bmatrix} i \\ 1 \end{bmatrix}$ are eigenvectors of $A = \begin{bmatrix} \cos\vartheta & \sin\vartheta \\ -\sin\vartheta & \cos\vartheta \end{bmatrix}$.

If $P = \begin{bmatrix} 1 & i \\ i & 1 \end{bmatrix}$ compute the product $P^{-1}AP$.

9.12 For each of the following matrices determine an invertible matrix P such that $P^{-1}AP$ is diagonal with the diagonal entries in increasing order of magnitude:

$$\begin{bmatrix} 0 & 1 & 0 \\ 1 & 0 & 1 \\ 0 & 1 & 0 \end{bmatrix}, \quad \begin{bmatrix} 3 & -1 & 0 \\ -1 & 3 & 0 \\ 0 & 0 & 1 \end{bmatrix}.$$

9.13 Consider the system of recurrence relations

$$u_{n+1} = \alpha u_n + \beta v_n$$

$$v_{n+1} = u_n.$$

If the matrix $\begin{bmatrix} \alpha & \beta \\ 1 & 0 \end{bmatrix}$ has distinct eigenvalues λ_1, λ_2 prove that

$$u_n = \frac{\lambda_1^n}{\lambda_1 - \lambda_2}(u_1 - \lambda_2 u_0) - \frac{\lambda_2^n}{\lambda_1 - \lambda_2}(u_1 - \lambda_1 u_0).$$

What is the situation when the eigenvalues are not distinct?

9.14 Determine the n-th power of the matrix

$$\begin{bmatrix} 2 & 2 & 0 \\ 1 & 2 & 1 \\ 1 & 2 & 1 \end{bmatrix}.$$

9.15 Solve the system of equations

$$x_{n+1} = 2x_n + 6y_n$$
$$y_{n+1} = 6x_n - 3y_n$$

given that $x_1 = 0$ and $y_1 = -1$.

9.16 Given the matrix

$$A = \begin{bmatrix} 6\frac{1}{2} & -2\frac{1}{2} & 2\frac{1}{2} \\ -2\frac{1}{2} & 6\frac{1}{2} & -2\frac{1}{2} \\ 0 & 0 & 4 \end{bmatrix},$$

find a matrix B such that $B^2 = A$.

9.17 Given $A \in \text{Mat}_{n \times n} \mathbb{R}$ let $k = \max |a_{ij}|$. Prove that, for all positive integers r,

$$|[A^r]_{ij}| \leqslant k^r n^{r-1}.$$

For every scalar β, associate with A the infinite series

$$S_\beta(A) \equiv I_n + \beta A + \beta^2 A^2 + \cdots + \beta^r A^r + \cdots .$$

We say that $S_\beta(A)$ converges if each of the series

$$\delta_{ij} + \beta[A]_{ij} + \beta^2[A^2]_{ij} + \cdots + \beta^r[A^r]_{ij} + \cdots$$

converges. Prove that

(1) $S_\beta(A)$ converges if $|\beta| < \dfrac{1}{nk}$;

(2) if $S_\beta(A)$ converges then $I_n - \beta A$ has an inverse which is the sum of the series.

Deduce that if A is a real $n \times n$ matrix and λ is an eigenvalue of A then

$$|\lambda| \leqslant n \max |a_{ij}|.$$

The Minimum Polynomial

In Chapter 9 we introduced the notions of *eigenvalue* and *eigenvector* of a matrix or of a linear mapping. There we concentrated our attention on showing the importance of these notions in solving particular problems. Here we shall take a closer algebraic look.

We begin by considering again the vector space $\mathrm{Mat}_{n \times n} F$ which, as we know, has the natural basis $\{E_{ij} ; \ i, j = 1, \dots, n\}$ and so is of dimension n^2. Thus, recalling Corollary 3 of Theorem 5.8, we have that every set of $n^2 + 1$ elements of $\mathrm{Mat}_{n \times n} F$ must be linearly dependent. In particular, given any $A \in \mathrm{Mat}_{n \times n} F$, the $n^2 + 1$ powers

$$A^0 = I_n, \ A, \ A^2, \ A^3, \ \dots, A^{n^2}$$

are linearly dependent and so there is a non-zero polynomial

$$p(X) = a_0 + a_1 X + a_2 X^2 + \cdots + A_{n^2} X^{n^2} \in F[X]$$

such that $p(A) = 0$. The same is of course true for any $f \in \mathrm{Lin}(V, V)$ where V is of dimension n; for, by Theorem 7.2, we have $\mathrm{Lin}(V, V) \simeq \mathrm{Mat}_{n \times n} F$.

But we can do better than this: there is in fact a polynomial $p(X)$ which is *of degree at most n* and such that $p(A) = 0$.

This is the celebrated **Cayley–Hamilton Theorem** which we shall now establish. Since we choose to work in $\mathrm{Mat}_{n \times n} F$, the proof that we shall give is considered 'elementary'. There are other (much more 'elegant') proofs which use $\mathrm{Lin}(V, V)$.

Recall that if $A \in \mathrm{Mat}_{n \times n} F$ then the characteristic polynomial of A is

$$\chi_A(\lambda) = \det(A - \lambda I_n)$$

and that $\chi_A(\lambda)$ is of degree n in the indeterminate λ.

Theorem 10.1

[Cayley–Hamilton] $\quad \chi_A(A) = 0$.

Proof

Let $B = A - \lambda I_n$ and

$$\chi_A(\lambda) = \det B = b_0 + b_1 \lambda + \cdots + b_n \lambda^n.$$

Consider the adjugate matrix adj B. By definition, this is an $n \times n$ matrix whose entries are polynomials in λ of degree at most $n - 1$, and so we have

$$\text{adj } B = B_0 + B_1\lambda + \cdots + B_{n-1}\lambda^{n-1}$$

for some $n \times n$ matrices $B_0, \ldots, B_{n-1}$. Recalling from Theorem 8.8 that $B \cdot \text{adj } B = (\det B)I_n$, we have

$$(\det B)I_n = B \cdot \text{adj } B = (A - \lambda I_n)\,\text{adj } B = A\,\text{adj } B - \lambda\,\text{adj } B,$$

i.e. we have the polynomial identity

$$b_0 I_n + b_1 I_n \lambda + \cdots + b_n I_n \lambda^n = AB_0 + \cdots + AB_{n-1}\lambda^{n-1} - B_0\lambda - \cdots - B_{n-1}\lambda^n.$$

Equating coefficients of like powers, we obtain

$$b_0 I_n = AB_0$$
$$b_1 I_n = AB_1 - B_0$$
$$\vdots$$
$$b_{n-1} I_n = AB_{n-1} - B_{n-2}$$
$$b_n I_n = -B_{n-1}.$$

Multiplying the first equation on the left by $A^0 = I_n$, the second by A, the third by A^2, and so on, we obtain

$$b_0 I_n = AB_0$$
$$b_1 A = A^2 B_1 - AB_0$$
$$\vdots$$
$$b_{n-1} A^{n-1} = A^n B_{n-1} - A^{n-1} B_{n-2}$$
$$b_n A^n = -A^n B_{n-1}.$$

Adding these equations together, we obtain $\chi_A(A) = 0$. $\square$

The Cayley–Hamilton Theorem is really quite remarkable, it being far from obvious that an $n \times n$ matrix should satisfy a polynomial equation of degree n.

Suppose now that k is the lowest degree for which a polynomial $p(X)$ exists such that $p(A) = 0$. Dividing $p(X)$ by its leading coefficient, we obtain a monic polynomial $m(X)$ of degree k which has A as a zero. Suppose that $m'(X)$ is another monic polynomial of degree k such that $m'(A) = 0$. Then $m(X) - m'(X)$ is a non-zero polynomial of degree less than k which has A as a zero. This contradicts the above assumption on k. Consequently, $m(X)$ is the unique monic polynomial of least degree having A as a zero. This leads to the following:

Definition

If $A \in \text{Mat}_{n \times n} F$ then the **minimum polynomial** of A is the monic polynomial $m_A(X)$ of least degree such that $m_A(A) = 0$.

Theorem 10.2

If $p(X)$ is a polynomial such that $p(A) = 0$ then the minimum polynomial $m_A(X)$ divides $p(X)$.

Proof

By euclidean division, there are polynomials $q(X)$, $r(X)$ such that

$$p(X) = m_A(X)q(X) + r(X)$$

with $r(X) = 0$ or $\deg r(X) < \deg m_A(X)$. Now by hypothesis $p(A) = 0$, and by definition $m_A(A) = 0$. Consequently, we have $r(A) = 0$. By the definition of $m_A(X)$ we cannot then have $\deg r(X) < \deg m_A(X)$, and so we must have $r(X) = 0$. It follows that $p(X) = m_A(X)q(X)$ and so $m_A(X)$ divides $p(X)$. □

Corollary

$m_A(X)$ divides $\chi_A(X)$. □

It is immediate from the above Corollary that every zero of $m_A(X)$ is a zero of $\chi_A(X)$. The converse is also true:

Theorem 10.3

$m_A(X)$ and $\chi_A(X)$ have the same zeros.

Proof

Observe that if λ is a zero of $\chi_A(X)$ then λ is an eigenvalue of A and so there is a non-zero $\mathbf{x} \in \text{Mat}_{n \times 1} F$ such that $A\mathbf{x} = \lambda\mathbf{x}$. Given any

$$h(X) = a_0 + a_1 X + \cdots + a_k X^k$$

we then have

$$h(A)\mathbf{x} = a_0\mathbf{x} + a_1 A\mathbf{x} + \cdots + a_k A^k \mathbf{x}$$
$$= a_0\mathbf{x} + a_1 \lambda\mathbf{x} + \cdots + a_k \lambda^k \mathbf{x}$$
$$= h(\lambda)\mathbf{x}$$

whence $h(\lambda)$ is an eigenvalue of $h(A)$. Thus $h(\lambda)$ is a zero of $\chi_{h(A)}(X)$.

Now take $h(X)$ to be $m_A(X)$. Then for every zero λ of $\chi_A(X)$ we have that $m_A(\lambda)$ is a zero of

$$\chi_{m_A(A)}(X) = \chi_0(X) = \det(-XI_n) = (-1)^n X^n.$$

Since the only zeros of this are 0, we have $m_A(\lambda) = 0$ so λ is a zero of $m_A(X)$. □

Example 10.1

The characteristic polynomial of

$$A = \begin{bmatrix} 1 & 0 & 1 \\ 0 & 2 & 1 \\ -1 & 0 & 3 \end{bmatrix}$$

is $\chi_A(X) = (X-2)^3$. Since $A - 2I_3 \neq 0$ and $(A - 2I_3)^2 \neq 0$, we have $m_A(X) = \chi_A(X)$.

Example 10.2

For the matrix

$$A = \begin{bmatrix} 5 & -6 & -6 \\ -1 & 4 & 2 \\ 3 & -6 & 4 \end{bmatrix}$$

we have $\chi_A(X) = (X - 1)(X - 2)^2$. By Theorem 10.3, the minimum polynomial is therefore either $(X - 1)(X - 2)^2$ or $(X - 1)(X - 2)$. Since $(A - I_3)(A - 2I_3) = 0$, it follows that $m_A(X) = (X - 1)(X - 2)$.

Theorem 10.4

A square matrix is invertible if and only if the constant term in its characteristic polynomial is not zero.

Proof

If A is invertible then, by Theorem 9.1, 0 is not an eigenvalue of A, and therefore 0 is not a zero of the characteristic polynomial. The constant term in the characteristic polynomial $\chi_A(\lambda)$ is then non-zero.

Conversely, suppose that the constant term of $\chi_A(\lambda)$ is non-zero. By Cayley–Hamilton we have $\chi_A(A) = 0$ which, by the hypothesis, can be written in the form $A\,p(A) = I_n$ for some polynomial p. Hence A is invertible. $\square$

Example 10.3

The matrix

$$A = \begin{bmatrix} 1 & 1 & 1 \\ 0 & 1 & 1 \\ 0 & 0 & 1 \end{bmatrix}$$

is such that $\chi_A(X) = (X - 1)^3$. Thus, applying the Cayley–Hamilton Theorem, we have that

$$0 = (A - I_3)^3 = A^3 - 3A^2 + 3A - I_3$$

whence we deduce that

$$A^{-1} = A^2 - 3A + 3I_3 = \begin{bmatrix} 1 & -1 & 0 \\ 0 & 1 & -1 \\ 0 & 0 & 1 \end{bmatrix}.$$

EXERCISES

10.1 If $A \in \text{Mat}_{n \times n} F$ is invertible and $\deg m_A(X) = p$, prove that A^{-1} is a linear combination of $I_n, A, A^2, \ldots, A^{p-1}$.

10.2 Determine the characteristic and minimum polynomials of each of the following matrices:

$$\begin{bmatrix} 1 & 2 & 3 \\ 0 & 1 & 2 \\ 0 & 0 & 1 \end{bmatrix}, \quad \begin{bmatrix} 1 & 1 & 0 \\ -1 & 1 & 1 \\ 0 & 1 & -1 \end{bmatrix}, \quad \begin{bmatrix} 0 & 0 & 2 \\ 1 & 0 & -1 \\ 0 & 1 & 1 \end{bmatrix}.$$

10.3 Prove that the constant term in the characteristic polynomial of A is $\det A$.

10.4 Determine the minimum polynomial of the rotation matrix

$$R_\vartheta = \begin{bmatrix} \cos \vartheta & \sin \vartheta \\ -\sin \vartheta & \cos \vartheta \end{bmatrix}.$$

Show that if ϑ is not an integer multiple of π then R_ϑ has no real eigenvalues.

The notion of characteristic polynomial can be defined for a linear mapping as follows.

Given a vector space V of dimension n over F and a linear mapping $f : V \to V$, let A be the matrix of f relative to some fixed ordered basis of V. Then the matrix of f relative to any other ordered basis is of the form $P^{-1}AP$ where P is the transition matrix from the new basis to the old basis (recall Theorem 7.5). Now the characteristic polynomial of $P^{-1}AP$ is

$$\det (P^{-1}AP - \lambda I_n) = \det \left[P^{-1}(A - \lambda I_n)P \right]$$
$$= \det P^{-1} \cdot \det (A - \lambda I_n) \cdot \det P$$
$$= \det (A - \lambda I_n),$$

i.e. we have

$$\chi_{P^{-1}AP}(\lambda) = \chi_A(\lambda).$$

It follows that the characteristic polynomial is independent of the choice of basis, so we can define the characteristic polynomial of f to be the characteristic polynomial of any matrix that represents f.

A similar definition applies to the notion of the minimum polynomial of a linear mapping, namely as the minimum polynomial of any matrix that represents the mapping.

EXERCISES

10.5 Determine the characteristic and minimum polynomial of the differentiation map $D : \mathbb{R}_n[X] \to \mathbb{R}_n[X]$.

10.6 Determine the minimum polynomial of the linear mapping $f : \mathbb{R}^2 \to \mathbb{R}^2$ given by

$$f(x, y) = (x + 4y, \tfrac{1}{2}x - y).$$

The full significance of the minimum polynomial is something that we cannot develop here, it being beyond the scope of the present text. The interested reader may care to consult more advanced texts, investigating particularly the notions of **invariant subspace** and **direct sum of subspaces**. To whet the appetite, let us say simply that if the minimum polynomial of f is

$$[p_1(X)]^{e_1}[p_2(X)]^{e_2} \cdots [p_k(X)]^{e_k}$$

where the $p_i(X)$ are distinct irreducible polynomials then each of the subspaces $V_i = \text{Ker}\,[p_i(f)]^{e_i}$ is f-invariant and bases of the V_i can be 'glued' together to form a basis of V. The matrix of f is then of the **block diagonal** form

$$\begin{bmatrix} A_1 & & & \\ & A_2 & & \\ & & \ddots & \\ & & & A_k \end{bmatrix}.$$

This result is known as the **Primary Decomposition Theorem**.

In the case where each $p_i(X)$ is linear (i.e. all of the eigenvalues of f lie in F), the matrix of f is a **Jordan matrix**. In this, each of the A_i is a **Jordan block**, i.e. is of the form

$$\begin{bmatrix} J_1 & & & \\ & J_2 & & \\ & & \ddots & \\ & & & J_t \end{bmatrix}.$$

in which each J_i is an **elementary Jordan matrix**, i.e is of the form

$$\begin{bmatrix} \lambda & 1 & & & \\ & \lambda & 1 & & \\ & & \lambda & 1 & \\ & & & \ddots & \ddots & \\ & & & & \lambda & 1 \\ & & & & & \lambda \end{bmatrix}.$$

If the characteristic and minimum polynomials are

$$\chi_f(X) = \prod_{i=1}^{k}(X - \lambda_i)^{d_i}, \quad m_f(X) = \prod_{i=1}^{k}(X - \lambda_i)^{e_i}$$

then in the Jordan form the eigenvalue λ_i appears d_i times in the diagonal, and the number of elementary Jordan matrices associated with λ_i is the geometric multiplicity of λ_i, with at least one being of size $e_i \times e_i$.

As a particular case of this, we obtain another solution to the diagonalisability problem:

Theorem 10.5

If V is a non-zero finite-dimensional vector space over a field F then a linear mapping $f : V \rightarrow V$ (respectively, a square matrix over F) is diagonalisable if and only if its minimum polynomial is a product of distinct linear factors. $\square$

Example 10.4

The matrix of Example 10.2 is diagonalisable.

EXERCISES

10.7 Consider the linear mapping $f : \mathbb{R}^3 \rightarrow \mathbb{R}^3$ given by
$$f(x, y, z) = (x + z, \, 2y + z, \, -x + 3z).$$
Prove that f is not diagonalisable.

SUPPLEMENTARY EXERCISES

10.8 Let A, B be square matrices over $\mathbb{C}$ and suppose that there exist rectangular matrices P, Q over $\mathbb{C}$ such that $A = PQ$ and $B = QP$.

If $h(X)$ is any polynomial with complex coefficients, prove that
$$Ah(A) = Ph(B)Q.$$
Hence show that $Am_B(A) = 0 = Bm_A(B)$. Deduce that one of the following holds:
$$m_A(X) = m_B(X), \quad m_A(X) = Xm_B(X), \quad m_B(X) = Xm_A(X).$$

10.9 Express the $r \times r$ matrix
$$\begin{bmatrix} 1 & 1 & \dots & 1 \\ 2 & 2 & \dots & 2 \\ \vdots & \vdots & \ddots & \vdots \\ r & r & \dots & r \end{bmatrix}$$
as the product of a column matrix and a row matrix. Hence find its minimum polynomial.

10.10 Let $f : \mathbb{C}_2[X] \rightarrow \mathbb{C}_2[X]$ be linear and such that
$$f(1) = -1 + 2X^2$$
$$f(1 + X) = 2 + 2X + 3X^2$$
$$f(1 + X - X^2) = 2 + 2X + 4X^2.$$
Find the eigenvalues and the minimum polynomial of f.

1.1 $\begin{bmatrix} 2 & 3 & 4 \\ 3 & 4 & 5 \\ 4 & 5 & 6 \end{bmatrix}$. **1.2** $\begin{bmatrix} 1 & 0 & 1 \\ 0 & 1 & 0 \\ 1 & 0 & 1 \end{bmatrix}$. **1.3** $\begin{bmatrix} 1 & -1 & 1 \\ -1 & 1 & -1 \\ 1 & -1 & 1 \end{bmatrix}$. **1.4** $\begin{bmatrix} 0 & 1 & 1 & \dots & 1 \\ -1 & 0 & 1 & \dots & 1 \\ -1 & -1 & 0 & \dots & 1 \\ \vdots & \vdots & \vdots & \ddots & \vdots \\ -1 & -1 & -1 & \dots & 0 \end{bmatrix}$.

1.5 $\begin{bmatrix} 1 & 2 & 3 & 4 & 5 & 6 \\ 2 & 2 & 6 & 4 & 10 & 6 \\ 3 & 6 & 3 & 12 & 15 & 6 \\ 4 & 4 & 12 & 4 & 20 & 12 \\ 5 & 10 & 15 & 20 & 5 & 30 \\ 6 & 6 & 6 & 12 & 30 & 6 \end{bmatrix}$, $\begin{bmatrix} 1 & 1 & 1 & 1 & 1 & 1 \\ 1 & 2 & 1 & 2 & 1 & 2 \\ 1 & 1 & 3 & 1 & 1 & 3 \\ 1 & 2 & 1 & 4 & 1 & 2 \\ 1 & 1 & 1 & 1 & 5 & 1 \\ 1 & 2 & 3 & 2 & 1 & 6 \end{bmatrix}$. **1.6** $\begin{bmatrix} a_{1n} & \dots & a_{12} & a_{11} \\ a_{2n} & \dots & a_{22} & a_{21} \\ \vdots & & \vdots & \vdots \\ a_{nn} & \dots & a_{n2} & a_{n1} \end{bmatrix}$.

1.7 Real numbers are 1×1 matrices. **1.8** Theorem 1.3. **1.9** $\begin{bmatrix} w-y & x-z & y-w \\ w-y & x-z & y-w \end{bmatrix}$.

1.10 $X = \frac{3}{4}A + \frac{15}{8}B$. **1.11** $X = A + 2B = \begin{bmatrix} 3 & 2 & 2 \\ 2 & 3 & 2 \\ 2 & 2 & 3 \end{bmatrix}$. **1.12** $\begin{bmatrix} 4 & 0 & 0 \\ 0 & 4 & 0 \\ 0 & 0 & 4 \end{bmatrix}$.

1.13 $\begin{bmatrix} 3 & 2 & 1 \\ 2 & 2 & 1 \\ 1 & 1 & 1 \end{bmatrix}$. **1.14** $\begin{bmatrix} 1 & 2 & 3 & 4 \\ 2 & 4 & 6 & 8 \\ 3 & 6 & 9 & 12 \\ 4 & 8 & 12 & 16 \end{bmatrix}$, [30]. **1.15** Each is $\begin{bmatrix} 3 & 45 & 9 \\ 11 & -11 & 17 \\ 7 & 17 & 13 \end{bmatrix}$.

1.16 The product is the 1×1 matrix $[t]$ where $t = ax^2 + 2hxy + by^2 + 2gx + 2fy + c$. Each of the equations can be written in the form $[x \ y \ 1]M \begin{bmatrix} x \\ y \\ 1 \end{bmatrix} = [0]$ with, respectively, M the matrices

$\begin{bmatrix} 1 & \frac{9}{2} & 4 \\ \frac{9}{2} & 1 & \frac{5}{2} \\ 4 & \frac{5}{2} & 2 \end{bmatrix}$, $\begin{bmatrix} \frac{1}{\alpha^2} & 0 & 0 \\ 0 & \frac{1}{\beta^2} & 0 \\ 0 & 0 & -1 \end{bmatrix}$, $\begin{bmatrix} 0 & \frac{1}{2} & 0 \\ \frac{1}{2} & 0 & 0 \\ 0 & 0 & -\alpha^2 \end{bmatrix}$, $\begin{bmatrix} 0 & 0 & -2\alpha \\ 0 & 1 & 0 \\ -2\alpha & 0 & 0 \end{bmatrix}$.

1.17 $A^2 = \begin{bmatrix} 0 & 0 & a^2 \\ 0 & 0 & 0 \\ 0 & 0 & 0 \end{bmatrix}$, $A^3 = 0$.

1.18 We have $AB = 0, A^2 = A, B^2 = -B$, and $(A+B)^2 = I_2$. On the other hand, $A^2 + 2AB + B^2 = \begin{bmatrix} 1 & 2 \\ 0 & 1 \end{bmatrix}$. Since $(A + B)^2 = I_2$ we have $(A + B)^3 = (A + B)(A + B)^2 = A + B$; and since $AB = 0$, $A^3 = A, B^3 = B$ we have $A^3 + 3A^2B + 3AB^2 + B^3 = A + B$.

1.19 We have $a_{ij} = 0$ and $b_{ij} = 0$ whenever $i \neq j$ and so $[AB]_{ij} = \sum_k a_{ik}b_{kj} = 0$ if $i \neq j$.

1.20 By Exercise 1.19 and induction, if A is diagonal then so is A^p.

1.21 If A and B commute then a simple inductive argument shows that A^m and B commute for every positive integer m. Fixing m, the same induction shows that B^n commutes with A^m for every positive integer n.

1.22 The proof is the same as for the binomial theorem $(x + y)^n = \sum_{r=0}^{n} \binom{n}{r} x^r y^{n-r}$ and is by induction, using properties of the binomial coefficients and the hypothesis that $xy = yx$.

1.23 $[A(B + C)]' = [AB]' + [AC]' = [B'A'] + [C'A']$.

1.24 The result is trivial for $n = 1$. Suppose it holds for n. Then $(A^{n+1})' = (A^n A)' = A'(A^n)' = A'(A')^n = (A')^{n+1}$,

1.25 If $AB = BA$ then $A'B' = (BA)' = (AB)' = B'A'$.

1.26 Using $a^2 + b^2 + c^2 = 1$ we have $A^2 = \begin{bmatrix} a^2 - 1 & ab & ac \\ ab & b^2 - 1 & bc \\ ac & bc & c^2 - 1 \end{bmatrix} = X'X - I_2$. Multiplying on the left by A then gives $A^3 = -A$. Finally, $A^4 = -A^2$ where A^2 is as above.

1.27 If A is both symmetric and skew-symmetric then $a_{ij} = a_{ji}$ and $a_{ij} = -a_{ji}$ whence $a_{ij} = 0$.

1.28 $A' = A$ and $B' = -B$. Thus

$$(AB + BA)' = B'A' + A'B' = -BA - AB = -(AB + BA)$$

and so $AB + BA$ is skew-symmetric. Similarly, $AB - BA$ is symmetric. Next, $(A^2)' = (A')^2 = A^2$ so A^2 is symmetric. Similarly, B^2 is symmetric. Finally,

$$(A^p B^q A^p)' = (A')^p (B')^q (A')^p = A^p (-B)^q A^p$$

and so $A^p B^q A^p$ is symmetric if q is even, and skew-symmetric if q is odd.

1.29 $A' = (xy' - yx')' = yx' - xy' = -A$. If $\mathbf{x} = \begin{bmatrix} x_1 \\ \vdots \\ x_n \end{bmatrix}$ and $\mathbf{y} = \begin{bmatrix} y_1 \\ \vdots \\ y_n \end{bmatrix}$ then $\mathbf{x'y} = \sum_{i=1}^{n} x_i y_i = \mathbf{y'x}$.

If now $\mathbf{x'x} = \mathbf{y'y} = [1]$ and $\mathbf{x'y} = \mathbf{y'x} = [k]$ then

$$A^2 = (\mathbf{xy'} - \mathbf{yx'})(\mathbf{xy'} - \mathbf{yx'}) = \mathbf{xy'xy'} - \mathbf{yx'xy'} - \mathbf{xy'yx'} + \mathbf{yx'yx'} = k\mathbf{xy'} - \mathbf{yy'} - \mathbf{xx'} + k\mathbf{yx'}$$

and hence

$$
\begin{aligned}
A^3 &= (k\mathbf{xy'} - \mathbf{yy'} - \mathbf{xx'} + k\mathbf{yx'})(\mathbf{xy'} - \mathbf{yx'}) \\
&= k\mathbf{xy'xy'} - \mathbf{yy'xy'} - \mathbf{xx'xy'} + k\mathbf{yx'xy'} - k\mathbf{xy'yx'} + \mathbf{yy'yx'} + \mathbf{xx'yx'} - k\mathbf{yx'yx'} \\
&= k^2\mathbf{xy'} - k\mathbf{yy'} - \mathbf{xy'} + k\mathbf{yy'} - k\mathbf{xx'} + \mathbf{yx'} + k\mathbf{xx'} - k^2\mathbf{yx'} \\
&= k^2(\mathbf{xy'} - \mathbf{yx'}) - \mathbf{xy'} + \mathbf{yx'} \\
&= (k^2 - 1)A.
\end{aligned}
$$

1.30 $[AB - BA]_{11} = a_{12}b_{21} - a_{21}b_{12}$ and $[AB - BA]_{22} = a_{21}b_{12} - a_{12}b_{21}$ so the sum of the diagonal elements is 0. If $E = \begin{bmatrix} a & b \\ c & -a \end{bmatrix}$ then $E^2 = \begin{bmatrix} a^2 + bc & 0 \\ 0 & a^2 + bc \end{bmatrix} = (a^2 + bc)I_2$.

For the last part, observe from the above that $(AB - BA)^2 = \lambda I_2$ and therefore commutes with every C.

1.31 Let $X = \begin{bmatrix} a & b \\ c & d \end{bmatrix}$. Then $X^2 = I_2$ if and only if

$$a^2 + bc = 1, \quad b(a + d) = 0, \quad c(a + d) = 0, \quad cb + d^2 = 1.$$

Suppose that $b = 0$. Then these equations reduce to $a^2 = 1, c(a + d) = 0, d^2 = 1$ from which we see that *either* $a = d = 1, c = 0$; *or* $a = d = -1, c = 0$; *or* $a = 1, d = -1, c$ arbitrary; *or* $a = -1, d = 1, c$ arbitrary.

If $b \neq 0$ then we must have $a + d = 0$ whence $d = -a$ and $c = (1 - a^2)/b$.

Thus the possibilities for X are

$$\begin{bmatrix} 1 & 0 \\ 0 & 1 \end{bmatrix}, \quad \begin{bmatrix} -1 & 0 \\ 0 & -1 \end{bmatrix}, \quad \begin{bmatrix} 1 & 0 \\ c & -1 \end{bmatrix}, \quad \begin{bmatrix} -1 & 0 \\ c & 1 \end{bmatrix}, \quad \begin{bmatrix} a & b \\ (1 - a^2)/b & -a \end{bmatrix}.$$

1.32 If (x, y) lies on the curve $y^2 - x^2 = 1$ then $A = \begin{bmatrix} x & y \\ -y & -x \end{bmatrix}$ is such that $A^2 = -I_2$, so there are infinitely many such matrices.

1.33 Simple calculations reveal that $A^4 = 0$ whence $A^n = 0$ for $n \geqslant 4$ and so

$$B = A - \tfrac{1}{2}A^2 + \tfrac{1}{3}A^3 = \begin{bmatrix} 0 & a & \tfrac{1}{2}a^2 & \tfrac{1}{3}a^3 \\ 0 & 0 & a & \tfrac{1}{2}a^2 \\ 0 & 0 & 0 & a \\ 0 & 0 & 0 & 0 \end{bmatrix}.$$

Likewise, $B^4 = 0$ and so $B^n = 0$ for $n \geqslant 4$. Thus

$$B + \tfrac{1}{2!}B^2 + \tfrac{1}{3!}B^3 = \begin{bmatrix} 0 & a & a^2 & a^3 \\ 0 & 0 & a & a^2 \\ 0 & 0 & 0 & a \\ 0 & 0 & 0 & 0 \end{bmatrix} = A.$$

1.34 The result follows from the standard formulae $\cos(\vartheta + \varphi) = \cos \vartheta \cos \varphi - \sin \vartheta \sin \varphi$ and $\sin(\vartheta + \varphi) = \sin \vartheta \cos \varphi + \cos \vartheta \sin \varphi$.

1.35 For the inductive step, use the previous exercise:

$$A^{n+1} = A^n A = \begin{bmatrix} \cos n\vartheta & \sin n\vartheta \\ -\sin n\vartheta & \cos n\vartheta \end{bmatrix} \begin{bmatrix} \cos \vartheta & \sin \vartheta \\ -\sin \vartheta & \cos \vartheta \end{bmatrix} = \begin{bmatrix} \cos(n + 1)\vartheta & \sin(n + 1)\vartheta \\ -\sin(n + 1)\vartheta & \cos(n + 1)\vartheta \end{bmatrix}.$$

1.36 Let B_n be the matrix on the right. Then clearly $B_1 = A$ so the result is true for $n = 1$. For the inductive step, observe that $B_n A = B_{n+1}$.

1.37 $[[AB]C] = (AB - BA)C - C(AB - BA) = ABC + CBA - BAC - CAB$; and similarly $[[BC]A] = BCA + ACB - CBA - ABC$, $[[CA]B] = CAB + BAC - ACB - BCA$. The first result follows by adding these expressions together. As for the second, we have $[(A + B)C] = (A + B)C - C(A + B) = AC - CA + BC - CB = [AC] + [BC]$. The third result follows by expanding as in the first.

Take $A = B = \begin{bmatrix} 0 & 1 \\ 1 & 0 \end{bmatrix}, C = \begin{bmatrix} 1 & 0 \\ 0 & 0 \end{bmatrix}$. Then $[[AB]C] = 0, [A[BC]] = \begin{bmatrix} 2 & 0 \\ 0 & -2 \end{bmatrix}$.

1.38 Substitute for y in the expression for x and compare with the expression for x in terms of z. We have $c_1 = a_1 b_1 + a_2 b_3$, etc.

2.1 If A is orthogonal then $AA' = I_n = A'A$. Since $A = (A')'$, we then have $A'(A')' = I_n = (A')'A'$ so that A' is orthogonal.

2.2 $(AB)'AB = B'A'AB = B'I_nB = I_n$ and similarly $AB(AB)' = I_n$.

2.3 See Example 4.3 of the text.

2.4 Let the top sheet be the (x, y)-plane and the bottom sheet the (x', y')-plane. If ϑ is the angle of anti-clockwise rotation of the top sheet, we have

$$\begin{bmatrix} \frac{5}{13} \\ \frac{12}{13} \end{bmatrix} = R_{-\vartheta} \begin{bmatrix} 1 \\ 0 \end{bmatrix} = \begin{bmatrix} \cos\vartheta & -\sin\vartheta \\ \sin\vartheta & \cos\vartheta \end{bmatrix} \begin{bmatrix} 1 \\ 0 \end{bmatrix} = \begin{bmatrix} \cos\vartheta \\ \sin\vartheta \end{bmatrix}$$

and so $\cos\vartheta = \frac{5}{13}$ and $\sin\vartheta = \frac{12}{13}$. The point (x', y') above which the point $(2, 3)$ lies is then given by $\begin{bmatrix} x' \\ y' \end{bmatrix} = \begin{bmatrix} \frac{5}{13} & -\frac{12}{13} \\ \frac{12}{13} & \frac{5}{13} \end{bmatrix} \begin{bmatrix} 2 \\ 3 \end{bmatrix}$, i.e. $(x', y') = (-2, 3)$.

2.5 Clearly $x' = x$ and $y' = -y$, so $\begin{bmatrix} x' \\ y' \end{bmatrix} = \begin{bmatrix} 1 & 0 \\ 0 & -1 \end{bmatrix} \begin{bmatrix} x \\ y \end{bmatrix}$. The other matrix is $\begin{bmatrix} -1 & 0 \\ 0 & 1 \end{bmatrix}$.

2.6 To obtain (x_L, y_L), rotate the axes through ϑ, take the reflection in the new x-axis, then rotate through $-\vartheta$. The matrix in question is $R_{-\vartheta} M R_\vartheta$ where M is the matrix of the previous exercise. A simple calculation shows that this product is $\begin{bmatrix} \cos 2\vartheta & \sin 2\vartheta \\ \sin 2\vartheta & -\cos 2\vartheta \end{bmatrix}$.

2.7 Rotate the axes through ϑ, project onto the new x-axis, then rotate the axes through $-\vartheta$. The required matrix is

$$\begin{bmatrix} \cos\vartheta & -\sin\vartheta \\ \sin\vartheta & \cos\vartheta \end{bmatrix} \begin{bmatrix} 1 & 0 \\ 0 & 0 \end{bmatrix} \begin{bmatrix} \cos\vartheta & \sin\vartheta \\ -\sin\vartheta & \cos\vartheta \end{bmatrix} = \begin{bmatrix} \cos^2\vartheta & \sin\vartheta\cos\vartheta \\ \sin\vartheta\cos\vartheta & \sin^2\vartheta \end{bmatrix}.$$

2.8 $A(p\mathbf{x}_1 + q\mathbf{x}_2) = pA\mathbf{x}_1 + (1-p)A\mathbf{x}_2 = p\mathbf{b} + (1-p)\mathbf{b} = \mathbf{b}$.

2.9 If $X = \begin{bmatrix} a & b \\ c & d \end{bmatrix}$ is such that $X^2 = 0$ then we have $a^2 + bc = 0$, $b(a+d) = 0$, $c(a+d) = 0$, $bc + d^2 = 0$. If $b = 0$ then clearly $a = d = 0$ and $X = \begin{bmatrix} 0 & 0 \\ c & 0 \end{bmatrix}$ which is of the required form.

If $b \neq 0$ then $d = -a$ and $a^2 + bc = 0$ which gives $X = \begin{bmatrix} a & b \\ -a^2/b & -a \end{bmatrix}$. Writing $z = \sqrt{b}$ and $w = a/\sqrt{b}$, we see that X is again of the required form. The result fails for real matrices since, for example, $A = \begin{bmatrix} 0 & 0 \\ 1 & 0 \end{bmatrix}$ is such that $A^2 = 0$ but there is no b such that $-b^2 = 1$.

2.10 $[\overline{A}']_{ij} = \overline{z_{ji}} = [\overline{A}']_{ij}$. That $\overline{A + B} = \overline{A} + \overline{B}$ and $\overline{AB} = \overline{B}\,\overline{A}$ follow immediately from similar properties of complex numbers.

2.11 $(A + \overline{A'})' = A' + (\overline{A'})' = A' + \overline{A} = \overline{A'} + A$. Hence $A + \overline{A'}$ is hermitian. Similarly, $A - \overline{A'}$ is skew-hermitian. For the last part, follow exactly the proof of Theorem 1.10.

3.1 ρ_3 becomes ρ_1; ρ_4 becomes ρ_2; ρ_1 becomes ρ_3; ρ_2 becomes ρ_4.

3.2 ρ_3 becomes ρ_1; ρ_1 becomes ρ_2; ρ_2 becomes ρ_3; ρ_4 remains the same.

3.3 ρ_1 becomes $\gamma\rho_3$; ρ_2 becomes $\delta\rho_4$; ρ_3 becomes $\alpha\rho_1$; ρ_4 becomes $\beta\rho_2$.

3.4 ρ_1 becomes $\alpha\rho_2 + \rho_3$; ρ_2 becomes $\beta\rho_2$; ρ_3 becomes $\rho_1 + \gamma\rho_2$.

3.5 $\begin{bmatrix} 1 & 2 & 3 \\ 3 & 1 & 2 \\ 5 & 5 & 8 \end{bmatrix} \rightsquigarrow \begin{bmatrix} 1 & 2 & 3 \\ 0 & -5 & -7 \\ 0 & -5 & -7 \end{bmatrix} \rightsquigarrow \begin{bmatrix} 1 & 2 & 3 \\ 0 & -5 & -7 \\ 0 & 0 & 0 \end{bmatrix}$; $\begin{bmatrix} 1 & 2 & 3 \\ 3 & 1 & 2 \\ 2 & 3 & 1 \end{bmatrix} \rightsquigarrow \begin{bmatrix} 1 & 2 & 3 \\ 0 & -5 & -7 \\ 0 & -1 & -5 \end{bmatrix} \rightsquigarrow \begin{bmatrix} 1 & 2 & 3 \\ 0 & -5 & -7 \\ 0 & 0 & -\frac{18}{5} \end{bmatrix}$.

3.6 $\begin{bmatrix} 1 & 2 & 0 & 3 & 1 \\ 1 & 2 & 3 & 3 & 3 \\ 1 & 0 & 1 & 1 & 3 \\ 1 & 1 & 1 & 2 & 1 \end{bmatrix} \rightsquigarrow \begin{bmatrix} 1 & 2 & 0 & 3 & 1 \\ 0 & 0 & 3 & 0 & 2 \\ 0 & -2 & 1 & -2 & 2 \\ 0 & -1 & 1 & -1 & 0 \end{bmatrix} \rightsquigarrow \begin{bmatrix} 1 & 2 & 0 & 3 & 1 \\ 0 & -1 & 1 & -1 & 0 \\ 0 & -2 & 1 & -2 & 2 \\ 0 & 0 & 3 & 0 & 2 \end{bmatrix} \rightsquigarrow$

$\begin{bmatrix} 1 & 2 & 0 & 3 & 1 \\ 0 & -1 & 1 & -1 & 0 \\ 0 & 0 & -1 & 0 & 2 \\ 0 & 0 & 3 & 0 & 2 \end{bmatrix} \rightsquigarrow \begin{bmatrix} 1 & 2 & 0 & 3 & 1 \\ 0 & -1 & 1 & -1 & 0 \\ 0 & 0 & -1 & 0 & 2 \\ 0 & 0 & 0 & 0 & 8 \end{bmatrix}$.

3.7 $\alpha = 1$. **3.8** $\begin{bmatrix} 1 & 0 & 0 & -2 & -3 & 0 \\ 0 & 1 & 0 & -1 & 2 & 1 \\ 0 & 0 & 1 & -1 & -5 & 2 \\ 0 & 0 & 0 & 0 & 0 & 0 \end{bmatrix}$. **3.9** Each is I_4. **3.10** Each is $\begin{bmatrix} 1 & 0 & 0 & 1 \\ 0 & 1 & 1 & 0 \\ 0 & 0 & 0 & 0 \\ 0 & 0 & 0 & 0 \end{bmatrix}$.

3.11 $\lambda_1 \rho_1 + \lambda_2 \rho_2 + \lambda_3 \rho_3 = 0$ gives $\lambda_1 + \lambda_3 = 0 = 2\lambda_1 + \lambda_2 = \lambda_2 + \lambda_3$, the only solution of which is $\lambda_1 = \lambda_2 = \lambda_3 = 0$.

3.12 The maximum number of independent rows (and columns) is 3.

3.13 The row rank is 2.

3.14 $\begin{bmatrix} 1 & 1 & 1 & 1 & 4 \\ 1 & \lambda & 1 & 1 & 4 \\ 1 & 1 & \lambda & 3-\lambda & 6 \\ 2 & 2 & 2 & \lambda & 6 \end{bmatrix} \rightsquigarrow \begin{bmatrix} 1 & 1 & 1 & 1 & 4 \\ 0 & \lambda-1 & 0 & 0 & 0 \\ 0 & 0 & \lambda-1 & 2-\lambda & 2 \\ 0 & 0 & 0 & \lambda-2 & -2 \end{bmatrix}$ so if $\lambda \neq 1, 2$ then the row rank is 4. When $\lambda = 1$ the row rank is 2, and when $\lambda = 2$ the row rank is 4.

3.15 The Hermite form of each matrix is I_3 so they are row-equivalent.

3.16 The row and column rank are each 3.

3.17 Proceed exactly as in Example 3.26. $P = \begin{bmatrix} \frac{1}{2} & \frac{1}{2} & 0 \\ -\frac{1}{4} & \frac{1}{4} & 0 \\ -2 & -3 & 1 \end{bmatrix}$ and $Q = \begin{bmatrix} 1 & 0 & 1 \\ 0 & -1 & 1 \\ 0 & 0 & 1 \end{bmatrix}$ do the trick. (The solution is not unique, so check your answer by direct multiplication.)

3.18 $\begin{bmatrix} I_n & 0 \\ 0 & 0 \end{bmatrix}$ for $n = 1, 2, 3, 4$.

3.19 The number of zero rows in the normal form of A is $n - \text{rank } A$ and the number of zero rows in the normal form of B is $m - \text{rank } B$. Since the latter must be less than the former we have $n - \text{rank } A \geqslant m - \text{rank } B$ whence $\text{rank } B \geqslant m - n + \text{rank } A$.

3.20 By Theorem 3.10, row-equivalent matrices have the same rank and so are equivalent.

3.21 Both A and A' have the same normal form. **3.22** Each has normal form $\begin{bmatrix} I_3 & 0 \end{bmatrix}$.

3.23 The coefficient matrix has rank 3 whereas the augmented matrix has rank 4. There is therefore no solution (Theorem 3.17).

3.24 A row-echelon form of the augmented matrix is

$$\begin{bmatrix} 1 & 1 & 0 & 1 & 4 \\ 0 & -2 & 0 & -6 & -1 \\ 0 & 0 & 1 & -1 & 1 \\ 0 & 0 & 0 & 0 & \alpha-1 \end{bmatrix}.$$

Thus the coefficient matrix and the augmented matrix have the same rank (i.e. the system has a solution) only when $\alpha = 1$. In this case, since the rank is 3 and the number of unknowns is 4 we can assign $4 - 3 = 1$ solution parameter. Taking this to be t, the general solution is $x = 2t + \frac{7}{2}$, $y = -3t + \frac{1}{2}, z = 1 + t$.

3.25 In the augmented matrix interchange the first two columns and the first two rows. This has no effect on the rank. A row-echelon form of the resulting matrix is

$$\begin{bmatrix} 1 & 2 & 1 & -6\alpha \\ 0 & \beta-6 & -1 & 20\alpha \\ 0 & 0 & \beta & 4+6\alpha \end{bmatrix}.$$

Thus if $\beta \neq 6,0$ the rank of the coefficient matrix is 3, as is that of the augmented matrix. Hence a unique solution exists (Theorem 3.18). When $\beta = 0$ the last line of the above matrix becomes $\begin{bmatrix} 0 & 0 & 0 & 4+6\alpha \end{bmatrix}$ so a solution exists only if $\alpha = -\frac{2}{3}$. In this case the solution is $x = -\frac{1}{6}z + \frac{20}{7}$ and $y = -\frac{2}{3}z - \frac{4}{9}$ where z is a parameter. When $\beta = 6$ the above matrix becomes

$$\begin{bmatrix} 1 & 2 & 1 & -6\alpha \\ 0 & 0 & -1 & 20\alpha \\ 0 & 0 & 6 & 4+6\alpha \end{bmatrix} \rightsquigarrow \begin{bmatrix} 1 & 2 & 1 & -6\alpha \\ 0 & 0 & -1 & 20\alpha \\ 0 & 0 & 0 & 4+126\alpha \end{bmatrix}.$$

In this case a solution exists if and only if $\alpha = -\frac{2}{63}$ and the general solution is $x = -\frac{1}{9}y - \frac{2}{9}$, $z = \frac{40}{63}$.

3.26 $\begin{bmatrix} 1 & 0 & 0 \\ 0 & 1 & 0 \\ 0 & 0 & -2 \end{bmatrix}$; $\begin{bmatrix} 1 & 0 & 0 \\ 0 & 0 & 1 \\ 0 & 1 & 0 \end{bmatrix}$; $\begin{bmatrix} 1 & -2 & 0 \\ 0 & 1 & 0 \\ 0 & 0 & 1 \end{bmatrix}$.

3.27 It is readily seen that rows 1, 3, 4 are linearly independent, so $r \geqslant 3$. If $r = 3$ then the second row must be a linear combination of rows 1, 3, 4. This is the case if and only if $a = 0 = d$ and $bc = 1$.

3.28 If $AX = B$ then necessarily X is of size 3×2. Let the columns of X be $\mathbf{x}_1$ and $\mathbf{x}_2$. Then $AX = B$ is equivalent to the two equations $A\mathbf{x}_1 = \mathbf{0}$ (homogeneous) and $A\mathbf{x}_2 = \mathbf{b}_2$ (non-homogeneous). By the usual reduction method the latter is easily seen to be consistent if and only if $\alpha = -1$.

3.29

$$\begin{bmatrix} 1 & -1 & 0 & -1 & -5 & \alpha \\ 2 & 1 & -1 & -4 & 1 & \beta \\ 1 & 1 & 1 & -4 & -6 & \gamma \\ 1 & 4 & 2 & -8 & -5 & \delta \end{bmatrix} \rightsquigarrow \begin{bmatrix} 1 & -1 & 0 & -1 & -5 & \alpha \\ 0 & 1 & -2 & 1 & 12 & \beta-\alpha-\gamma \\ 0 & 0 & 5 & -5 & -25 & \alpha-2\beta+3\gamma \\ 0 & 0 & 0 & 0 & 0 & 8\alpha-\beta-11\gamma+5\delta \end{bmatrix}$$

so a solution exists if and only if $8\alpha - \beta - 11\gamma + 5\delta = 0$, and in this case the rank of the matrix is 3. When $\alpha = \beta = -1, \gamma = 3, \delta = 8$ the solution is $x = 2u+3t, y = 1+u-2t, z = 2+u+5t$ where u, t are parameters.

3.30 Interchange the second and third columns, and the first and second rows. The matrix becomes

$$\begin{bmatrix} 1 & \lambda & 0 & 1 \\ -2 & -\lambda & \mu+3 & -3 \\ 2 & 3\lambda & 4 & -\lambda \end{bmatrix} \rightsquigarrow \begin{bmatrix} 1 & \lambda & 0 & 1 \\ 0 & \lambda & \mu+3 & -1 \\ 0 & 0 & 1-\mu & -\lambda-1 \end{bmatrix}.$$

If $\mu \neq 1$ then the rank is 3 and a unique solution exists. If $\mu = 1$ then a solution exists if and only if $\lambda = -1$. In this case the rank is 2 and the general solution involves one parameter.

3.31 $\min\{m,n\}$.　　**3.32** $\begin{bmatrix} 1 & -1 \\ -1 & 1 \end{bmatrix}$.

3.33 Consider $A = \begin{bmatrix} 1 & -1 \\ -1 & 1 \end{bmatrix}$ and $B = \begin{bmatrix} 1 & 0 \\ 0 & 0 \end{bmatrix}$. The normal form of A is B so these matrices are equivalent. But they are clearly not row-equivalent.

3.34 Let A and B be equivalent. Then A can be transformed into normal form N by a sequence of row and column operations, and then N can be transformed into B by a sequence of row and column operations (namely the 'inverses' of those that transform B into N). Hence A can be transformed into B by a sequence of row and column operations. Conversely, if A can be obtained from B by a sequence of row and column operations then A and B must have the same normal form.

3.35 $a = 0, b = 1, c = 2$.

3.36 Treat the system as a system of equations in which all calculations are done modulo 5. For example, adding the first row to the second gives $\begin{bmatrix} 0 & 2 & 1 & 2 & 2 & 2 \end{bmatrix}$.

4.1 The given matrix has rank 2. By Theorem 4.1, it therefore has neither a left inverse nor a right inverse.

4.2 Proceed as in Example 4.3:

$$\begin{bmatrix} 1 & 1 & 1 \\ 1 & 2 & 3 \\ 0 & 1 & 1 \end{bmatrix}^{-1} = \begin{bmatrix} 1 & 0 & -1 \\ 1 & -1 & 2 \\ -1 & 1 & -1 \end{bmatrix}; \qquad \begin{bmatrix} 1 & 2 & 1 \\ 1 & 3 & 2 \\ 1 & 0 & 1 \end{bmatrix}^{-1} = \begin{bmatrix} \frac{3}{2} & -1 & \frac{1}{2} \\ \frac{1}{2} & 0 & -\frac{1}{2} \\ -\frac{3}{2} & 1 & \frac{1}{2} \end{bmatrix};$$

$$\begin{bmatrix} 1 & 2 & 2 \\ 1 & 3 & 1 \\ 1 & 1 & 3 \end{bmatrix} \text{ has rank 2 so is not invertible;}$$

$$\begin{bmatrix} 1 & 1 & 1 & 1 \\ 1 & 2 & -1 & 2 \\ 1 & -1 & 2 & 1 \\ 1 & 3 & 3 & 2 \end{bmatrix}^{-1} = \begin{bmatrix} \frac{7}{3} & -\frac{1}{3} & -\frac{1}{3} & -\frac{2}{3} \\ \frac{4}{9} & -\frac{1}{9} & -\frac{4}{9} & \frac{1}{9} \\ -\frac{1}{9} & -\frac{2}{9} & \frac{1}{9} & \frac{2}{9} \\ -\frac{5}{3} & \frac{2}{3} & \frac{2}{3} & \frac{1}{3} \end{bmatrix}; \qquad \begin{bmatrix} 1 & 1 & 1 & 1 \\ 1 & 3 & 1 & 2 \\ 1 & 2 & -1 & 1 \\ 5 & 9 & 1 & 6 \end{bmatrix} \text{ has rank 3 so is not invertible.}$$

4.3 The product $A\mathbf{x}$ can be written as $x_1\mathbf{a}_1 + x_2\mathbf{a}_2 + \cdots + x_n\mathbf{a}_n$, i.e. as a linear combination of the columns of A. Then $A\mathbf{x} = \mathbf{0}$ has only the trivial solution if and only if $\mathbf{a}_1, \ldots, \mathbf{a}_n$ are linearly independent, which is the case if and only if A has rank n, which is so if and only if A is invertible.

4.4 If $c = 0$ we have $A = \begin{bmatrix} a & b \\ 0 & d \end{bmatrix}$ which is of rank 2 (i.e. A is invertible) if and only if $a \neq 0$, $d \neq 0$ which is equivalent to $ad \neq 0$. If $a = 0$ then $A \rightsquigarrow \begin{bmatrix} c & d \\ 0 & b \end{bmatrix}$ which is of rank 2 if and only if $bc \neq 0$. If $a \neq 0$ and $c \neq 0$ then $\begin{bmatrix} a & b \\ c & d \end{bmatrix} \rightsquigarrow \begin{bmatrix} a & b \\ 0 & ad - bc \end{bmatrix} a\rho_2 - c\rho_1$ which is of rank 2 if and only if $ad - bc \neq 0$. In this case we have $A^{-1} = \frac{1}{ad-bc} \begin{bmatrix} d & -b \\ -c & a \end{bmatrix}$.

4.5 $\begin{bmatrix} 1 & 1 & 0 \\ 1 & 0 & 0 \\ 1 & 2 & \alpha \end{bmatrix} \rightsquigarrow \begin{bmatrix} 1 & 0 & 0 \\ 0 & 1 & 0 \\ 0 & 0 & \alpha \end{bmatrix}$ which is of rank 3 if and only if $\alpha \neq 0$. In this case, by the process of Example 4.3, $A^{-1} = \begin{bmatrix} 0 & 1 & 0 \\ 1 & -1 & 0 \\ -\frac{2}{\alpha} & \frac{1}{\alpha} & \frac{1}{\alpha} \end{bmatrix}$.

4.6 The proof is by induction. The result is trivial for $p = 1$. By Theorem 4.4, the inductive step is $(A_1 \cdots A_p A_{p+1})^{-1} = A_{p+1}^{-1}(A_1 \cdots A_p)^{-1} = A_{p+1}^{-1} A_p^{-1} \cdots A_1^{-1}$.

4.7 That $A^3 = 0$ is routine. Using this fact, we have

$$
\begin{aligned}
A_x A_y &= (I_3 + xA + \tfrac{1}{2}x^2 A^2)(I_3 + yA + \tfrac{1}{2}y^2 A^2)\\
&= I_3 + xA + \tfrac{1}{2}x^2 A^2 + yA + xyA^2 + \tfrac{1}{2}y^2 A^2\\
&= I_3 + (x+y)A + \tfrac{1}{2}(x+y)^2 A^2\\
&= A_{x+y}.
\end{aligned}
$$

It follows that $A_x A_{-x} = A_0 = I_3$ whence A_x is invertible with $A_x^{-1} = A_{-x}$.

4.8 That $A_n A_m = A_{n+m}$ is routine. It follows that $A_n A_{-n} = A_0 = I_2$ so A_n is invertible with $A_n^{-1} = A_{-n}$. Similarly, $B_n B_m = B_{n+m}$ and $B_n^{-1} = B_{-n}$.

Finally, by Theorem 4.4, $(A_n B_m)^{-1} = B_{-m} A_{-n}$.

4.9 Since $(I_n + A)(I_n - A) = I_n - A^2 = (I_n - A)(I_n + A)$ we have that

$$(I_n + A)(I_n - A)(I_n + A)^{-1} = (I_n - A)(I_n + A)(I_n + A)^{-1} = I_n - A$$

and so $(I_n - A)(I_n + A)^{-1} = (I_n + A)^{-1}(I_n - A)$.

If A is skew-symmetric then $A' = -A$ and we have

$$
\begin{aligned}
PP' &= (I_n - A)(I_n + A)^{-1}[(I_n - A)(I_n + A)^{-1}]'\\
&= (I_n + A)^{-1}(I_n - A)(I_n + A')^{-1}(I_n - A')\\
&= (I_n + A)^{-1}(I_n - A)(I_n - A)^{-1}(I_n + A)\\
&= I_n.
\end{aligned}
$$

It follows that P is invertible with $P^{-1} = P'$, i.e. P is orthogonal.

For the given matrix A it is readily seen that

$$
I_n - A = \begin{bmatrix} 1 & -\cos\vartheta & 0 \\ \cos\vartheta & 1 & -\sin\vartheta \\ 0 & \sin\vartheta & 1 \end{bmatrix}, \quad (I_n + A)^{-1} = \begin{bmatrix} \tfrac{1}{2}(1 + \sin^2\vartheta) & -\tfrac{1}{2}\cos\vartheta & \tfrac{1}{2}\sin\vartheta\cos\vartheta \\ \tfrac{1}{2}\cos\vartheta & \tfrac{1}{2} & -\tfrac{1}{2}\sin\vartheta \\ \tfrac{1}{2}\sin\vartheta\cos\vartheta & \tfrac{1}{2}\sin\vartheta & \tfrac{1}{2}(1 + \cos^2\vartheta) \end{bmatrix}
$$

whence P is the matrix stated.

4.10
$$
\begin{bmatrix}
1 & -1 & 0 & -1 \\
0 & -\tfrac{1}{2} & 0 & 0 \\
-\tfrac{1}{5} & 1 & \tfrac{1}{5} & \tfrac{3}{5} \\
\tfrac{2}{5} & -\tfrac{1}{2} & -\tfrac{2}{5} & -\tfrac{1}{5}
\end{bmatrix}.
$$

4.11 If B is row-equivalent to A then there is an elementary matrix P such that $B = PA$. Thus, if A is invertible, we have that B is a product of invertible matrices and so is also invertible.

4.12 If AB is invertible then there exist elementary matrices P, Q such that $PABQ = I_n$. It follows that PA and QB are invertible, and from $(PA)^{-1}PA = I_n$ and $BQ(BQ)^{-1} = I_n$ we see that A and B are invertible.

4.13 $(A + B)A^{-1}(A - B) = (I_n + BA^{-1})(A - B) = A - B + B - BA^{-1}B = A - BA^{-1}B$, and similarly $(A - B)A^{-1}(A + B) = A - BA^{-1}B$ whence we have the required equality.

4.14 Expand the right-hand product using the distributive law. The resulting sum is the left-hand side. If now $A^s = 0$ then $A^{s+1} = 0$ and the equality gives

$$I_n = (I_n - A)(I_n + A + A^2 + \cdots + A^{s-1})$$

whence $I_n - A$ is invertible.

For the last part, follow the instructions; $A^{-1} = \begin{bmatrix} 0 & -1 & 0 & 0 \\ 1 & 0 & 2 & 1 \\ 1 & -1 & 3 & 1 \\ 0 & -1 & 1 & 1 \end{bmatrix}$.

4.15 We have

$$(BA - I_n)\big[B(AB - I_n)^{-1}A - I_n\big]$$

$$= \underbrace{BAB}(AB - I_n)^{-1}A - BA - B(AB - I_n)^{-1}A + I_n$$

$$= \underbrace{[B(AB - I_n) + B]}(AB - I_n)^{-1}A - BA - B(AB - I_n)^{-1}A + I_n$$

$$= BA + B(AB - I_n)^{-1}A - BA - B(AB - I_n)^{-1}A + I_n$$

$$= I_n.$$

Hence $BA - I_n$ is also invertible.

5.2 (1) and (2) are subspaces; (3) is not since it does not contain $(0, 0, 0, 0)$; (4) is not a subspace since, for example, $(1, -1, 1, -1)$ and $(0, 0, 0, 1)$ belong to the set but their sum $(1, -1, 1, 0)$ does not.

5.3 The sum of two continuous functions is continuous, and every scalar multiple of a continuous function is continuous.

5.4 The sum of two differentiable functions is differentiable, and every scalar multiple of a differentiable function is differentiable.

5.5 (1) If A and B are symmetric $n \times n$ then so is $A + B$; and so is λA for every λ. Hence the set of symmetric $n \times n$ matrices is a subspace.

(2) The set of invertible $n \times n$ matrices is not a subspace since every subspace of $\mathrm{Mat}_{n \times n} \mathbb{R}$ must contain the zero matrix, and this is not invertible.

(3) The matrices $\begin{bmatrix} 1 & 0 \\ 0 & 0 \end{bmatrix}$ and $\begin{bmatrix} 0 & 0 \\ 0 & 1 \end{bmatrix}$ are not invertible, but their sum is I_2 which is invertible. Hence the set of non-invertible matrices is not a subspace.

5.6 If $Ax = 0$ and $Ay = 0$ then $A(x + y) = Ax + Ay = 0$, and $A(\lambda x) = \lambda Ax = \lambda 0 = 0$. Thus the solutions of $Ax = 0$ form a subspace of $\mathrm{Mat}_{m \times 1} \mathbb{R}$.

5.7 Every subspace must contain the zero of the parent space.

5.8 The rank of the coefficient matrix is 3 and so for all given a, b, c the system is consistent.

Thus, for every $\begin{bmatrix} a \\ b \\ c \end{bmatrix} \in \mathrm{Mat}_{3 \times 1} \mathbb{R}$ there exist scalars $\lambda_1, \lambda_2, \lambda_3$ such that

$$\begin{bmatrix} a \\ b \\ c \end{bmatrix} = \lambda_1 \begin{bmatrix} 1 \\ 1 \\ 1 \end{bmatrix} + \lambda_2 \begin{bmatrix} 1 \\ 2 \\ 3 \end{bmatrix} + \lambda_3 \begin{bmatrix} 1 \\ 3 \\ 2 \end{bmatrix}$$

whence the three column matrices span $\mathrm{Mat}_{3 \times 1} \mathbb{R}$.

5.9 No; for example, the constant polynomial 1 cannot be expressed as a linear combination of the two given polynomials.

5.10 Let $E_1 = \begin{bmatrix} 1 & 1 \\ 0 & 0 \end{bmatrix}$, $E_2 = \begin{bmatrix} 0 & 0 \\ 1 & 1 \end{bmatrix}$, $E_3 = \begin{bmatrix} 1 & 0 \\ 0 & 1 \end{bmatrix}$, $E_4 = \begin{bmatrix} 0 & 1 \\ 1 & 1 \end{bmatrix}$. Then we have

$$\begin{bmatrix} a & b \\ c & d \end{bmatrix} = xE_1 + yE_2 + zE_3 + tE_4$$

if and only if $x + z = 0$, $x + t = 0$, $y + t = 0$, $y + z + t = 0$. The coefficient matrix of this system of equations is

$$A = \begin{bmatrix} 1 & 0 & 1 & 0 \\ 1 & 0 & 0 & 1 \\ 0 & 1 & 0 & 1 \\ 0 & 1 & 1 & 1 \end{bmatrix}$$

which is of rank 4 (hence invertible). The system therefore has a solution (which is in fact unique). Hence the given set is a spanning set.

5.11 (1) Suppose that $\sum\limits_{i=1}^{k} \lambda_i x_i = 0$ where each $x_i \in S_1$. Since $S_1 \subseteq S_2$ we have that each $x_i \in S_2$ and so, since S_2 is linearly independent (by hypothesis), every $\lambda_i = 0$. Hence S_1 is linearly independent.

(2) If S_1 is linearly dependent then by Theorem 5.5 at least one element of S_1 can be expressed as a linear combination of other elements in S_1. But $S_1 \subseteq S_2$, so all of these elements belong to S_2. By Theorem 5.5 again, therefore, S_2 is linearly dependent.

5.12 The sets (1) and (2) are linearly independent since the 3×3 matrices formed from them are each of rank 3. As for (3), this set is linearly dependent; the third column matrix is the sum of the first two.

5.13 (1) Linearly independent; consider the entries in the $(2,1)$ position.

(2) Linearly independent; take a linear combination of the four matrices to be equal to the zero matrix and solve the corresponding equations (only the zero solution possible).

(3) Linearly dependent; we have

$$\begin{bmatrix} 2 & 3 \\ 4 & 3 \end{bmatrix} = \tfrac{1}{2}\begin{bmatrix} 1 & 0 \\ 0 & 2 \end{bmatrix} + \tfrac{3}{2}\begin{bmatrix} 1 & 1 \\ 2 & 1 \end{bmatrix} + \tfrac{1}{2}\begin{bmatrix} 0 & 3 \\ 2 & 1 \end{bmatrix}.$$

5.14 Follow the process in Example 5.22. (1) and (2) are linearly independent. (3) is linearly dependent; we have

$$13 + X = 3(1 + X + 2X^2) - 2(-5 + X + 3X^2).$$

5.15 Every $p(X) \in \mathbb{R}_n[X]$ can be written uniquely in the form

$$p(X) = a_0 + a_1 X + a_2 X^2 + \cdots + a_n X^n.$$

5.16 Every $A \in \mathrm{Mat}_{m \times n} \mathbb{R}$ can be written uniquely in the form $A = \sum\limits_{p=1}^{m} \sum\limits_{q=1}^{n} a_{pq} E_{pq}$.

5.17 The sum of two diagonal matrices is a digonal matrix, and every scalar multiple of a diagonal matrix is a diagonal matrix. Hence the diagonal matrices form a subspace. A basis for this subspace is the set of diagonal matrices E_{pp} of the previous exercise.

5.18 The set of Toeplitz matrices is clearly closed under addition and multiplication by scalars, and so forms a subspace. A basis consists of the Toeplitz matrices E_{pq} where $p \neq q$ and the Toeplitz matrix I_n.

5.19 Since $\cos 2x = \cos^2 x - \sin^2 x$ we have that $W = \mathrm{Span}\{f, g\}$. Now $f = \cos^2$ and $g = \sin^2$ are linearly independent. To see this, let

$$\lambda_1 \cos^2 x + \lambda_2 \sin^2 x = 0.$$

Differentiate to get $(\lambda_1 - \lambda_2) \sin x \cos x = 0$. Since this must hold for all x we must have $\lambda_1 = \lambda_2$, and since the original equation holds for all x this means that $\lambda_1 = \lambda_2 = 0$. Hence a basis for W is $\{\cos^2, \sin^2\}$.

5.20 Apply the process of Example 5.25. (1) is a basis; (2) is not.

5.21 Apply the process of Example 5.25. The matrix

$$A = \begin{bmatrix} 1 & -1 & 1 & 0 \\ 1 & -1 & 1 & 1 \\ 0 & 1 & 1 & -1 \\ 0 & 2 & 3 & -3 \end{bmatrix}$$

is invertible so the given set is a basis. Since

$$A^{-1} = \begin{bmatrix} 2 & -1 & 5 & -2 \\ 0 & 0 & 3 & -1 \\ -1 & 1 & -2 & 1 \\ -1 & 1 & 0 & 0 \end{bmatrix}$$

we deduce that

$$(a,b,c,d) = \lambda_1(1,1,0,0) + \lambda_2(-1,-1,1,2) + \lambda_3(1,-1,1,3) + \lambda_4(0,1,-1,-3)$$

where $\lambda_1 = 2a - b + 5c - d$, $\lambda_2 = 3c - d$, $\lambda_3 = -a + b - 2c + d$, $\lambda_4 = -a + b$.

5.22 Writing $X = \{x_1, x_2\}$ and $Y = \{y_1, y_2\}$ we observe that $y_1 = \frac{1}{2}(x_2 - x_1)$ and $y_2 = 2x_2 - x_1$. It follows that $Y \subseteq \operatorname{Span} X$ and therefore $\operatorname{Span} Y \subseteq \operatorname{Span} X$. Similarly, $x_1 = y_2 - 4y_1$ and $x_2 = y_2 - 2y_1$ whence we have the reverse inclusion $\operatorname{Span} X \subseteq \operatorname{Span} Y$.

5.23 (1) $\Rightarrow$ (2) : Since $\{v_1, v_2\}$ is a basis we can write w_1 and w_2 as unique linear combinations of w_1, w_2 and so there is a matrix A such that $\begin{bmatrix} w_1 \\ w_2 \end{bmatrix} = A \begin{bmatrix} v_1 \\ v_2 \end{bmatrix}$. Since $\operatorname{Span}\{w_1, w_2\} = V$, we can write v_1, v_2 as linear combinations of w_1, w_2 and so there is a matrix B such that $\begin{bmatrix} v_1 \\ v_2 \end{bmatrix} = B \begin{bmatrix} w_1 \\ w_2 \end{bmatrix}$. Consequently we have $\begin{bmatrix} v_1 \\ v_2 \end{bmatrix} = BA \begin{bmatrix} v_1 \\ v_2 \end{bmatrix}$. Since $\{v_1, v_2\}$ is a basis we must have $BA = I_2$ whence A is invertible.

(2) $\Rightarrow$ (1) : If $\begin{bmatrix} w_1 \\ w_2 \end{bmatrix} = A \begin{bmatrix} v_1 \\ v_2 \end{bmatrix}$ with A invertible then we have $\begin{bmatrix} v_1 \\ v_2 \end{bmatrix} = A^{-1} \begin{bmatrix} w_1 \\ w_2 \end{bmatrix}$ whence $\{v_1, v_2\} \subseteq \operatorname{Span}\{w_1, w_2\}$. Since $\{v_1, v_2\}$ is a basis of V it follows that $V = \operatorname{Span}\{w_1, w_2\}$.

5.24 Suppose that $x \in A \cap B$. Then $x = \lambda_1 a_1 + \lambda_2 a_2$ where a_1, a_2 are the elements of the given spanning set of A, and similarly $x = \lambda_3 b_1 + \lambda_4 b_2$ where b_1, b_2 are the elements of the given spanning set of B. Consequently we have $\lambda_1 a_1 + \lambda_2 a_2 - \lambda_3 b_1 - \lambda_4 b_2 = 0$. But, as can readily be verified, the elements a_1, a_2, b_1, b_2 are linearly independent. Hence each $\lambda_i = 0$ and consequently $x = 0$. Thus we see that $A \cap B = \{0\}$ and so has dimension 0.

5.25 If $\{v_1, \ldots, v_n\}$ is a basis of V as a vector space over $\mathbb{C}$ then every $x \in V$ can be written uniquely as $x = \sum_{k=1}^{n} \alpha_k v_k$ where each $\alpha_k \in \mathbb{C}$. Writing $\alpha_k = a_k + ib_k$ where $a_k, b_k \in \mathbb{R}$ we have $x = \sum_{k=1}^{n} a_k v_k + \sum_{k=1}^{n} b_k(iv_k)$. Thus $\{v_1, \ldots, v_n, iv_1, \ldots, iv_n\}$ is a spanning set of V over $\mathbb{R}$. This spanning set is linearly independent, for if $\sum_{k=1}^{n} a_k v_k + \sum_{k=1}^{n} b_k(iv_k) = 0$ then $\sum_{k=1}^{n} a_k v_k = 0$ and $i\sum_{k=1}^{n} b_k v_k = 0$ whence $a_k = 0 = b_k$ for each k since $\{v_1, \ldots, v_n\}$ is a basis over $\mathbb{C}$. Hence the above set is a basis for V over $\mathbb{R}$ in which case V is a real vector space of dimension $2n$.

5.26 We have that $(a, b, c, d) \in \text{Span } X$ if and only if the system

$$\begin{bmatrix} 2 & 7 & 3 & 2 \\ 2 & 5 & 2 & 1 \\ 1 & 5 & 2 & 2 \\ 3 & 5 & 1 & 1 \end{bmatrix} \begin{bmatrix} x \\ y \\ z \\ t \end{bmatrix} = \begin{bmatrix} a \\ b \\ c \\ d \end{bmatrix}$$

is consistent. Now, by the usual reduction method,

$$\begin{bmatrix} 2 & 7 & 3 & 2 & a \\ 2 & 5 & 2 & 1 & b \\ 1 & 5 & 2 & 2 & c \\ 3 & 5 & 1 & 1 & d \end{bmatrix} \rightsquigarrow \begin{bmatrix} 1 & 5 & 2 & 2 & c \\ 0 & 1 & 0 & 1 & -2a+b+2c \\ 0 & 0 & 2 & -2 & 10a-6b-8c \\ 0 & 0 & 0 & 0 & -10a+10b+6c-2d \end{bmatrix}.$$

For consistency, we therefore require $5a - 5b - 3c + d = 0$.

The first three vectors of X form a basis for Span X. Since $\mathbb{R}^4$ is of dimension 4, any vector not dependent on these three vectors (i.e. any vector not satisfying the above condition) may be added to obtain a basis for $\mathbb{R}^4$, e.g. the vector $(0, 1, 1, 0)$.

5.27 As a third basis vector, $\begin{bmatrix} 0 \\ 0 \\ 1 \end{bmatrix}$ will do.

5.28 Recall from Exercise 5.15 that $\mathbb{R}_n[X]$ is of dimension $n + 1$. A possible basis is obtained by adding the monomials X^2 and X^3.

5.29 (1) is true. The standard conditions are $x + y \in W$ and $\lambda x \in W$. These together imply $\lambda x + \mu y \in W$. Conversely, if the latter holds, take $\lambda = \mu = 1$, then $\mu = 0$, to obtain the former.

(2) is false; the subspace has dimension 1 with basis $\{(1, 1, 1)\}$.

(3) is true. If the given spanning set is also linearly independent then it forms a basis and there is nothing to prove. If not, then at least one element is a linear combination of others. Removing this element, we still have a spanning set. Continuing in this way we discard elements one by one, obtaining a smaller spanning set each time. Do this until the remaining set is also linearly independent (in the worst case this will have only one element); it will then form a basis.

(4a) is true. We have $a(1, 2, 1) + b(2, 2, 1) = (a + 2b, 2a + 2b, a + b)$. Taking $a + b = y$ and $b = x$ we see that so also is (4b).

(5a) is true. We can extend a basis of P to a basis of Q.

(5b) is false. For example, consider $P = \{(x, x, x) ; x \in \mathbb{R}\}$, $Q = \{(x, y, 0) ; x, y \in \mathbb{R}\}$. We have dim $P = 1$ (see (2) above) and dim $Q = 2$, but $P \nsubseteq Q$.

(6) is false. For example, take $w = -x$ and recall that 0 cannot belong to a basis.

5.30 (1) Yes. (2) No. For example $(1, 0, 0, 0)$ and $(0, 1, 0, 0)$ belong to the set but their sum does not. (3) Yes. The set is $\{(0, 0, c, d) ; c, d \in \mathbb{R}\}$. (4) No. For example, $(1, 0, 0, 0)$ and $(0, 1, 0, 0)$ belong to the set but their sum does not.

5.31 Both sets are closed under addition and multiplication by scalars, so are subspaces.

5.32 Given $\alpha, \beta, \gamma \in \mathbb{C}$, to determine $\lambda, \mu, \nu \in \mathbb{C}$ such that

$$(\alpha, \beta, \gamma) = \lambda(3 - i, 2 + 2i, 4) + \mu(2, 2 + 4i, 3) + \nu(1 - i, -2i, -1),$$

solve the resulting equations for λ, μ, ν to obtain the unique solution

$$\begin{bmatrix} \lambda \\ \mu \\ \nu \end{bmatrix} = -\frac{1}{12 + 12i} \begin{bmatrix} -2+2i & 5-3i & -6-6i \\ 2-6i & -7+5i & 6+6i \\ -2-10i & -1+-3i & 6+6i \end{bmatrix} \begin{bmatrix} \alpha \\ \beta \\ \gamma \end{bmatrix}.$$

This shows that the given set is a basis. Moreover, taking $\alpha = 1$, $\beta = \gamma = 0$ we obtain $(1, 0, 0)$ in terms of the basis, and similarly for the others.

5.33 We have

$$\begin{bmatrix} 1 & 2 & 2 & 1 & -1 \\ 0 & 2 & 2 & -1 & -2 \\ 2 & 6 & 2 & 1 & -4 \\ 1 & 4 & 0 & 0 & -3 \end{bmatrix} \rightsquigarrow \begin{bmatrix} 1 & 0 & 0 & 2 & 1 \\ 0 & 1 & 0 & -\frac{1}{2} & -1 \\ 0 & 0 & 1 & 0 & 0 \\ 0 & 0 & 0 & 0 & 0 \end{bmatrix}.$$

The general solution is therefore

$$\mathbf{x} = \begin{bmatrix} -2t - w \\ \frac{1}{2}t + w \\ 0 \\ t \\ w \end{bmatrix} = t \begin{bmatrix} -2 \\ \frac{1}{2} \\ 0 \\ 1 \\ 0 \end{bmatrix} + w \begin{bmatrix} -1 \\ 1 \\ 0 \\ 0 \\ 1 \end{bmatrix}.$$

The solution space is therefore of dimension 2.

5.34 As defined, $A + B$ is closed under addition and multiplication by scalars, so is a subspace. If C is a subspace such that $A \subseteq C$ and $B \subseteq C$ then for $a \in A$ and $b \in B$ we have $a, b \in C$ and so $a + b \in C$. Hence $A + B \subseteq C$.

To establish the equality we show that LHS$\supseteq$RHS and LHS$\subseteq$RHS. For the former, observe that LHS $\supseteq L \cap M$ and LHS $\supseteq L \cap N$ whence, by the above, LHS$\supseteq$RHS. As for the latter, if $x \in$ RHS then $x = y + z$ where $y \in L \cap M$ and $z \in L \cap N$. Since $y, z \in L$ we have $x = y + z \in L$. Moreover, since $y \in M$ we have $x = y + z \in M + (L \cap N)$. It follows that $x \in L \cap [M + (L \cap M)]$ =LHS.

For the last part, take

$$L = \{(x, x, 0) \, ; \, x, y \in \mathbb{R}\}, \quad M = \{(0, y, z) \, ; \, y, z \in \mathbb{R}\}, \quad N = \{(x, 0, 0) \, ; \, x \in \mathbb{R}\}.$$

Then $M + N = \mathbb{R}^3$, $L \cap (M + N) = L$, $L \cap M = \{(0, 0, 0)\}$, $L \cap N = \{(0, 0, 0)\}$, and the stated inequality holds.

5.35 It is readily verified that E_n is closed under addition and multiplication by scalars and so is a subspace of Map$(\mathbb{R}, \mathbb{R})$.

Suppose now that f is the zero map in E_1. Then we have

$$(\forall x \in \mathbb{R}) \qquad a_0 + a_1 \cos x + b_1 \sin x = 0.$$

Taking $x = 0$ we obtain $a_0 + a_1 = 0$, and taking $x = \pi/2$ we obtain $a_0 + b_1 = 0$. Thus $a_1 = b_1 = -a_0$. Taking $x = \pi/4$ we obtain $a_0 + \frac{1}{\sqrt{2}}a_1 + \frac{1}{\sqrt{2}}b_1 = 0$ whence $a_0 = a_1 = b_1 = 0$. Suppose now, by way of induction, that the zero map of E_{n-1} (with $n \geqslant 2$) has all its coefficients 0 and let f be the zero map of E_n. It is readily verified that $D^2f + n^2f$ is given by the prescription

$$(D^2f + n^2f)(x) = n^2 a_0 + \sum_{k=1}^{n-1} (n^2 - k^2)(a_k \cos kx + b_k \sin kx)$$

and since f is the zero map of E_n we have that $D^2f + n^2f$ isthe zero map of E_{n-1}. By the induction hypothesis, therefore, we have that all the coefficients $a_0, a_1, \ldots, a_{n-1}, b_1, \ldots, b_{n-1}$ are 0 and the formula for f reduces to

$$(\forall x \in \mathbb{R}) \qquad 0 = f(x) = a_n \cos x + b_n \sin x.$$

Taking $x = 0$ we obtain $a_n = 0$, and taking $x = \pi/2n$ we obtain $b_n = 0$. Thus all the coefficients of f are 0 and the result follows by induction.

It is clear that the $2n + 1$ functions generate E_n. Moreover, by what we have just proved, the only linear combination of these $2n + 1$ functions that is zero is the trivial linear combination. Hence these functions constitute a basis for E_n.

5.36 Let M be the set of rational functions described. Then clearly M is closed under addition and multiplication by scalars, so is a subspace of $\text{Map}(\mathbb{R}, \mathbb{R})$.

Following the hint, observe that each $f \in M$ can be written uniquely in the form

$$f = a_0 f_0 + \cdots + a_{r+s-1} f_{r+s-1}$$

and so the f_i form a basis of M.

For the last part, it suffices by the hint (and by Corollary 2 of Theorem 5.8) to show that B is linearly independent. For this purpose, suppose that

$$\frac{a_1}{x - \alpha} + \cdots + \frac{a_r}{(x - \alpha)^r} + \frac{b_1}{x - \beta} + \cdots + \frac{b_s}{(x - \beta)^s} = 0.$$

Multiplying both sides by $(x - \alpha)^r (x - \beta)^s$ we obtain

$$a_1(x - \alpha)^{r-1}(x - \beta)^s + a_2(x - \alpha)^{r-2}(x - \beta)^s + \cdots + a_r(x - \beta)^s +$$

$$b_1(x - \alpha)^r (x - \beta)^{s-1} + b_2(x - \alpha)^r (x - \beta)^{s-2} + \cdots + b_r(x - \alpha)^r = 0.$$

Taking the term $a_r(x - \beta)^s$ over to the RHS, what remains on the LHS is divisible by $x - \alpha$ and, since $\alpha \neq \beta$, we deduce that $a_r = 0$. Similarly, we see that $b_s = 0$. Extracting a resultant factor $(x - \alpha)(x - \beta)$, we can repeat this argument to obtain $a_{r-1} = 0 = b_{s-1}$. Continuing in this way, we see that every coefficient is 0 and therefore that B is linearly independent. Hence B is a basis.

5.37 The result is trivial if $n = 1$ since f_1 is non-zero. By way of induction, suppose that $\{f_1, \ldots, f_{n-1}\}$ is linearly independent whenever $r_1, \ldots, r_{n-1}$ are distinct. Consider $\{f_1, \ldots, f_n\}$ and suppose that $r_1, \ldots, r_n$ are distinct. If $\lambda_1 f_1 + \cdots + \lambda_n f_n = 0$ then

$$(\forall x \in \mathbb{R}) \qquad \lambda_1 e^{r_1 x} + \cdots + \lambda_n e^{r_n x} = 0.$$

Dividing by $e^{r_n x}$ (which is non-zero) and differentiating, we obtain

$$\lambda_1(r_1 - r_n)e^{(r_1 - r_n)x} + \cdots + \lambda_{n-1}(r_{n-1} - r_n)e^{(r_{n-1} - r_n)x} = 0.$$

Since the $n - 1$ real numbers $r_1 - r_n, \ldots, r_{n-1} - r_n$ are distinct, the induction hypothesis shows that $\lambda_1 = \cdots = \lambda_{n-1} = 0$. Consequently, $\lambda_n f_n = 0$ and hence $\lambda_n = 0$ (since $e^{r_n x} \neq 0$). Thus $f_1, \ldots, f_n$ are linearly independent. Hence the result by induction. Conversely, if the r_i are not distinct then $r_i = r_j$ for some i, j whence $f_i = f_j$ and the f_i are dependent.

5.38 Suppose that $\sum_{i=0}^{n} \lambda_i P_i(X) = 0$. Since $\deg p_i(X) = i$ we have, on differentiating n times, $\lambda_n = 0$ whence $\sum_{i=0}^{n-1} \lambda_i P_i(X) = 0$. Differentiating $n - 1$ times, we deduce that $\lambda_{n-1} = 0$. Continuing in this way we see that every coefficient $\lambda_i = 0$ and therefore the given set is linearly independent. Since $\dim \mathbb{R}[X] = n + 1$ and there are $n + 1$ such functions, they therefore form a basis (recall Corollary 2 of Theorem 5.8).

5.39 Since sums and scalar multiples of step functions are also step functions it is clear that the set E is a subspace of the real vector space of all mappings from $\mathbb{R}$ to $\mathbb{R}$. Given $\vartheta \in E$, the step function ϑ_i that agrees with ϑ on the interval $[a_i, a_{i+1}[$ and is zero elsewhere is given by the prescription

$$\vartheta_i(x) = \vartheta(a_i)\big[e_{a_i}(x) - e_{a_{i+1}}(x)\big].$$

Since then $\vartheta = \sum_{i=0}^{n+1} \vartheta_i$ it follows that $\{e_k \; ; \; k \in [0, 1[\}$ spans E. Since the functions e_k are linearly independent they therefore form a basis of E.

Similarly, the set F of piecewise linear functions is a vector space.

That G is a subspace is clear since it is closed under addition and multiplication by scalars. That $\{g_k \; ; \; k \in [0, 1[\}$ is a basis of G is similar to the above.

That every $f \in F$ can be expressed uniquely in the form $g + e$ where $g \in G$ and $e \in E$ can be seen from geometric considerations. A typical $f \in F$ can be depicted as in the diagram

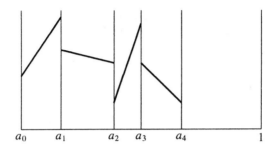

Now think of strips of wood that can slide up and down and thereby manufacture g via e.

6.1 (1), (2), and (5) are linear; (3) is not linear since $f(1, 0, 0) + f(-1, 0\,0) \neq f(0, 0, 0)$; (4) and (6) are not linear since in each case, for example, $f[2(0, 0, 0)] \neq 2f(0, 0, 0)$.

6.2 (1) and (2) are linear. (3) is not linear since in general $T_B(\lambda I_n) \neq \lambda T_B(I_n)$.

6.3 (1) and (2) are linear; (3) is not linear since in general $f(\lambda p(X)) \neq \lambda f(p(X))$.

6.4 $f_A(\mathbf{x} + \mathbf{y}) = A(\mathbf{x} + \mathbf{y}) = A\mathbf{x} + A\mathbf{y} = f_A(\mathbf{x}) + f_A(\mathbf{y})$ and $f_A(\lambda \mathbf{x}) = A\lambda \mathbf{x} = \lambda A\mathbf{x} = \lambda f_A(\mathbf{x})$.

6.5 By theorems in analysis, $\int (p + q) = \int p + \int q$ and $\int \lambda p = \lambda \int p$.

6.6 No. A and $-A$ have the same Hermite form, so $f(A) = f(-A)$ whence $f(A) - f(-A) = 0$. But $f\big(A - (-A)\big) = f(2A) = f(A)$ since $2A$ has the same Hermite form as A.

6.7 Observe that $T(0) = 7B^2$ so if T is linear we must have $B^2 = 0$. Conversely, if $B^2 = 0$ then $T(A) = AB + BA - 2BA - 3AB = -2AB - BA$ which is linear.

6.8 $D^{\rightarrow}(\mathbb{R}_n[X]) = \mathbb{R}_{n-1}[X]$ and $D^{\leftarrow}(\{0\}) = \mathbb{R}$ (the set of constant polynomials).

6.9 We have that $f^{\rightarrow}f^{\leftarrow}f^{\rightarrow}(A) = \{f(b) \; ; \; b \in f^{\leftarrow}f^{\rightarrow}(A)\} = \{f(b) \; ; \; f(b) \in f^{\rightarrow}(A)\} = \{f(b) \; ; \; f(b) = f(a) \text{ for some } a \in A\} = f^{\rightarrow}(A)$, and $f^{\leftarrow}f^{\rightarrow}f^{\leftarrow}(B) = \{a \; ; \; f(a) \in f^{\rightarrow}f^{\leftarrow}(B)\} = \{a \; ; \; f(a) = f(x) \text{ where } f(x) \in B\} = \{a \; ; \; f(a) \in B\} = f^{\leftarrow}(B)$.

6.10 $D^{\rightarrow}(X)$ is the set of polynomials whose even coefficients are zero. $D^{\leftarrow}(X)$ is the set of polynomials whose even coefficients, except possibly the constant term, are zero.

6.11 The image of the x-axis is the line $y = x$; the image of the y-axis is the line $y = -x$; and the image of the line $y = mx$ where $m \neq 1$ is the line $y = \frac{1+m}{1-m} x$.

6.12 Observe that if $x \in f^{\leftarrow}(Y)$ then $f(x) \in Y$ and therefore we have $f^{\rightarrow}\big(f^{\leftarrow}(Y)\big) \subseteq Y$. Since $f^{\rightarrow}$ is inclusion-preserving, it follows that $f^{\rightarrow}[X \cap f^{\leftarrow}(Y)] \subseteq f^{\rightarrow}(X) \cap f^{\rightarrow}\big(f^{\leftarrow}(Y)\big) \subseteq f^{\rightarrow}(X) \cap Y$. For the reverse inclusion observe that if $a \in f^{\rightarrow}(X) \cap Y$ then there exists $b \in X$ such that $f(b) = a \in Y$ whence $b \in x \cap f^{\leftarrow}(Y)$ and so $a = f(b) \in f^{\rightarrow}[X \cap f^{\leftarrow}(Y)]$.

6.13 $f(a, b, c) = a(1, 0, 1) + b(1, 1, 0) + c(0, 1, 1)$. $\operatorname{Im} f = \operatorname{Span}\{(1, 0, 1), (1, 1, 0), (0, 1, 1)\} = \mathbb{R}^3$, and $\operatorname{Ker} f = \{(0, 0, 0)\}$.

6.14 $\operatorname{Ker} f = \{(a, b) \in \mathbb{R}^2 \; ; \; b = 0\} = \operatorname{Im} f$.

6.15 $f : \mathbb{R}^3 \to \mathbb{R}^3$ given by $f(a, b, c) = (c, 0, 0)$ is such that $\operatorname{Im} f \subset \operatorname{Ker} f$; and $g : \mathbb{R}^3 \to \mathbb{R}^3$ given by $g(a, b, c) = (b, c, 0)$ is such that $\operatorname{Ker} g \subset \operatorname{Im} g$.

6.16 $f(a, b, c, d, e) = a(1, 1, 2, 0) + c(-1, 0, -1, -1) + d(3, 2, 5, 1) + e(-1, -1, -1, 0)$ and so

Im f is spanned by the rows of the matrix

$$\begin{bmatrix} 1 & 1 & 2 & 0 \\ -1 & 0 & -1 & -1 \\ 3 & 2 & 5 & 1 \\ -1 & -1 & -1 & 0 \\ 0 & 0 & 0 & 0 \end{bmatrix} \quad \text{which has Hermite form} \quad \begin{bmatrix} 1 & 0 & 0 & 1 \\ 0 & 1 & 0 & -1 \\ 0 & 0 & 1 & 0 \\ 0 & 0 & 0 & 0 \\ 0 & 0 & 0 & 0 \end{bmatrix}.$$

A basis of Im f is $\{(1,1,2,0),(-1,0,-1,-1),(-1,-1,-1,0)\}$. It is readily seen that Ker $f = \{(-2c,b,c,0)\ ;\ b,c \in \mathbb{R}\}$, so a basis for Ker f is $\{(0,1,0,0,0),(-2,0,1,1,0)\}$.

6.17 A basis of Im f is $\{X^2, X^3\}$; Ker $f = \mathbb{R}$.

6.18 Take $f(x,y,z,t) = (x,0,z,0)$ and $g(x,y,z,t) = (0,y,0,t)$,

6.19 Argue as in Example 6.12, noting that the matrix $\begin{bmatrix} 1 & 1 & 1 \\ 2 & -1 & -1 \\ 1 & 2 & -1 \end{bmatrix}$ is invertible.

6.20 Suppose that $\sum_{i=1}^{n} a_i f(v_i) = 0$. Then $f\left(\sum_{i=1}^{n} a_i v_i\right) = 0$ whence, f being injective, $\sum_{i=1}^{n} a_i v_i = 0$. The fact that $a_1, \ldots, a_n$ are linearly independent gives each $a_i = 0$ whence $f(v_1), \ldots, f(v_n)$ are linearly independent.

6.21 (1) $\Rightarrow$ (2) : If (1) holds then for every $x \in V$ we have $f(x) \in \text{Im } f = \text{Ker } f$ whence $f^2(x) = 0$, so that $f^2 = 0$. If $f = 0$ then Ker $f = V$ whence the contradiction $n = \dim V = \dim \text{Ker } f = \dim \text{Im } f = 0$. Hence $f \neq 0$. By Theorem 6.4 we have $n = \dim V = \dim \text{Ker } f + \dim \text{Im } f = 2 \dim \text{Im } f$ so n is even and the rank of f is $\frac{1}{2}n$.

(2) $\Rightarrow$ (1) : If (2) holds then $f^2(x) = 0$ gives $f(x) \in \text{Ker } f$ whence Im $f \subseteq \text{Ker } f$. By the Dimension Theorem, $\dim V = \text{rank} + \text{nullity}$, i.e. $n = \frac{1}{2}n + \text{nullity}$. Hence $\dim \text{Ker } f = \frac{1}{2}n = \dim \text{Im } f$. It follows by Theorem 5.9 that Ker $f = \text{Im } f$.

6.22 Consider the differentiation map $D : \mathbb{R}^2[X] \to \mathbb{R}^2[X]$. It is not possible to write X^2 as the sum of an element of Im D and an element of Ker D.

6.23 Writing $f(x) = a \sin x + b \cos x$ we have

(1) $\vartheta(f) = 2a$ so $f \in \text{Ker } \vartheta$ if and only if $a = 0$. Hence Ker $\vartheta = \text{Span}\{\cos\}$ and the nullity is 1.

(2) $\vartheta(f) = 0$ so Ker $\vartheta = W$ and the nullity is 2.

(3) $\vartheta(f) = 0$ if and only if $a = 0$ in which case the nullity is 1.

6.24 If $\vartheta = af + bg + ch$ then $T(\vartheta)(x) = -cx$ so rank $T = 1$. Ker $T = \{af + bg\ ;\ a,b \in \mathbb{R}\}$ so the nullity of T is 2.

6.25 Take $f\left(\sum_{i=0}^{n} a_i X^i\right) = (a_0, \ldots, a_n)$.

6.26 If $(x,y) \in \text{Ker } \vartheta$ then $\left(x, y - f(x)\right) = (0,0)$ and so $x = 0$ and $y = f(x) = f(0) = 0$. Hence Ker $\vartheta = \{(0,0)\}$ and so ϑ is injective. By Theorem 6.5, ϑ is an isomorphism.

6.27 Recall Example 5.25: the matrix $\begin{bmatrix} 1 & 1 & 1 \\ 1 & 2 & 3 \\ 1 & 1 & 2 \end{bmatrix}$ is invertible, so $\{(1,1,1),(1,2,3),(1,1,2)\}$ is a basis of $\mathbb{R}^3$. Proceed as in Example 6.15 to obtain

$$f(x,y,z) = (4x - 4y + z, 5x - 5y + z, 8x - 10y + 3z).$$

6.28 $f(a + bX + cX^2) = af(1) + bf(X) + cf(X^2) = a + cX + bX^2 + cX^3$.

6.29 If $\mathbb{C}$ is considered as a complex vector space, then for $\lambda = 1 - i$ we have $f\left(\lambda(1 + i)\right) = f(2) = 2$ whereas $\lambda f(1 + i) = (1 - i)(1 - i) = -2i$, so that f is not linear.

6.30 f is linear if and only if

$$f(\lambda_1 v_1 + \lambda_2 v_2 + \lambda_3 v_3) = \lambda_1 f(v_1) + \lambda_2 f(v_2) + \lambda_3 f(v_3),$$

i.e. if and only if

$$(\lambda_1 + \mu)w_1 + (\lambda_2 + \lambda_3)w_2 = \lambda_1(1 + \mu)w_1 + \lambda_2(\mu w_1 + w_2) + \lambda_3(\mu w_1 + w_2).$$

This is the case if and only if $\mu = 0$. Then $f(xv_1 + yv_2 + zv_3) = 0$ if and only if $x = 0, y + z = 0$ and so $\operatorname{Ker} f = \{a(v_2 - v_3) \; ; \; a \in F\}$. Thus a basis of $\operatorname{Ker} f$ is $\{v_2 - v_3\}$.

6.31 A basis of $\operatorname{Im} f$ is $\{(1,0,0),(1,0,1)\}$ and a basis of $\operatorname{Ker} f$ is $\{(1,-1,-1)\}$. $f^{\leftarrow}(A) = \{(x,-x,z) \; ; \; x, z \in \mathbb{R}\}$, a basis of which is $\{(1,-1,0),(0,0,1)\}$.

6.32 $f(x,y,z) = 2(x-y)X + (y+z)X^2 + (x+z)X^3$. A basis of $\operatorname{Im} f$ is $\{2X + X^3, -2X + X^2\}$. A basis of $\operatorname{Ker} f$ is $\{(1,1,-1)\}$. This can be extended to the basis $\{(1,1,-1),(1,0,0),(1,1,0)\}$.

6.33 The matrix $\begin{bmatrix} -1 & 1 & 1 \\ 1 & -1 & 1 \\ 1 & 1 & -1 \end{bmatrix}$ is invertible so $\{a,b,c\}$ is a basis. We have

$$(x,y,z) = \tfrac{1}{2}(y+z)a + \tfrac{1}{2}(x+z)b + \tfrac{1}{2}(x+y)c$$

and so

$$f(x,y,z) = \left(y + \tfrac{1}{2}(x+z), \tfrac{1}{2}(z-y), \tfrac{1}{2}[(\lambda-1)x + (\lambda+1)y], \tfrac{1}{2}[(\lambda-1)y - x + \lambda z]\right).$$

f is injective if $\lambda \neq -1$ and $\lambda \neq 2$. The dimension of W is 2. If $\lambda = 2$ then $f(1,1,0) = \left(\tfrac{3}{2}, -\tfrac{1}{2}, 2, 0\right)$ and $f^{\leftarrow}\{(1,1,0,0)\} = \left(2, -\tfrac{2}{3}, \tfrac{4}{3}\right)$.

6.34 If $x = f(s_1)$ and $y = f(s_2)$ where $s_1, s_2 \in S$ then for $t \in [0,1]$ we have

$$tx + (1-t)y = f[ts_1 + (1-t)s_2] \in f^{\rightarrow}(S)$$

and so $f^{\rightarrow}(S)$ is also convex.

6.35 By the Dimension Theorem (6.4) we have

$$\sum_{i=1}^{n}(-1)^i \dim V_i = -\dim \operatorname{Im} f_1 - \dim \operatorname{Ker} f_1$$
$$+ \dim \operatorname{Im} f_2 + \dim \operatorname{Ker} f_2$$
$$- \dim \operatorname{Im} f_3 - \dim \operatorname{Ker} f_3$$
$$\vdots$$
$$+ (-1)^n \dim \operatorname{Im} f_n + (-1)^n \dim \operatorname{Ker} f_n$$

Since the sequence is exact, $\dim \operatorname{Ker} f_1 = 0$ and $\operatorname{Im} f_n = V_{n+1}$. Moreover, $\operatorname{Im} f_i = \operatorname{Ker} f_{i+1}$. The above display therefore reduces to $\sum_{i=1}^{n}(-1)^i \dim V_i = (-1)^n \dim V_{n+1}$ whence the result follows.

6.36 The rank is 2 so the nullity is 1.

6.37 (1) $(g \circ f)(x) = g(y)$ where $y = f(x)$ so $\operatorname{Im}(g \circ f) \subseteq \operatorname{Im} g$.

(2) If $f(x) = 0$ then $g[f(x)] = g(0) = 0$ so $\operatorname{Ker} f \subseteq \operatorname{Ker}(g \circ f)$.

(3) By (1) we have $\operatorname{rank}(g \circ f) \leqslant \operatorname{rank} g$. By (2) and Theorem 6.4, we have

$$\operatorname{rank}(g \circ f) = \dim V - \dim \operatorname{Ker}(g \circ f) \leqslant \dim V - \dim \operatorname{Ker} f = \operatorname{rank} f.$$

Hence $\operatorname{rank}(g \circ f) \leqslant \min\{\operatorname{rank} f, \operatorname{rank} g\}$.

Let $\{e_1, \ldots, e_n\}$ be a basis of V. Let $m = \dim W$ and $s = \operatorname{rank} f$. As basis for W we can take $\{f(e_1), \ldots, f(e_s), w_{s+1}, \ldots, w_m\}$. Then

$$\operatorname{Im} g = \operatorname{Span}\{gf(e_1), \ldots, gf(e_s), g(w_{s+1}), \ldots, g(w_m)\}$$
$$= \operatorname{Span}\{gf(e_1), \ldots, gf(e_p), g(w_{s+1}), \ldots, g(w_m)\}$$

where $p = \operatorname{rank} gf$. It follows that

$$\operatorname{rank} g \leqslant p + m - s = \operatorname{rank} gf + \dim W - \operatorname{rank} f.$$

6.38 $\{1, X, X^2, X^3\}$ is a basis of $\mathbb{R}_3[X]$ and $f(1) = 0, f(X) = 0, f(X^2) = 0, f(X^3) = 6(X - 1)$. So the nullity of f is 3 and the rank is $4 - 3 = 1$.

6.39 If $\{e_1, e_2, e_3\}$ is the natural basis of $\mathbb{R}^3$ then $f_y(e_1) = (0, -y_3, y_2), f_y(e_2) = (y_3, 0, -y_1)$, $f_y(e_3) = (-y_2, y_1, 0)$. Since $y \neq 0$ we can assume that $y_1 \neq 0$. Then we have

$$f_y(e_1) = -\frac{y_2}{y_1} f_y(e_2) - \frac{y_3}{y_1} f_y(e_3)$$

where $f_y(e_2), f_y(e_3)$ are linearly independent. Hence a basis for Im f_y is $\{f_y(e_2), f_y(e_3)\}$ so rank $f_y = 2$. It follows that dim Ker $f_y = 3 - 2 = 1$ whence Ker $f_y = \text{Span}\{y\}$ since $f_y(y) = 0$.

6.40 $f(1,0,0,0) = \begin{bmatrix} 1 & 0 \\ 0 & -1 \end{bmatrix}, f(0,1,0,0) = \begin{bmatrix} 0 & 1 \\ 1 & 0 \end{bmatrix}, f(0,0,1,0) = \begin{bmatrix} 0 & i \\ -i & 0 \end{bmatrix}, f(0,0,0,1) = \begin{bmatrix} 1 & 0 \\ 0 & 1 \end{bmatrix}$. So f carries a basis to a basis and therefore, by Theorem 6.5, is an isomorphism.

7.1 $\begin{bmatrix} 1 & 2 \\ 2 & -1 \\ -1 & 0 \end{bmatrix};$ $\begin{bmatrix} 1 & -2 \\ -3 & -2 \\ 2 & 3 \end{bmatrix}.$ **7.2** $\begin{bmatrix} 2 & -1 & 0 \\ 0 & -2 & -1 \end{bmatrix};$ $\begin{bmatrix} 0 & 2 & -1 \\ 1 & -1 & 0 \end{bmatrix}.$ **7.3** $\begin{bmatrix} 2 & 2 & 1 \\ 0 & 3 & 0 \\ 0 & -4 & 1 \end{bmatrix}.$

7.4 $f(0,1,0) = f(1,1,0) - f(1,0,0) = (2,-2,6)$ and $f(0,0,1) = f(1,1,1) - f(1,1,0) = (1,-2,3)$ and so the matrix of f relative to the natural basis of $\mathbb{R}^3$ is $\begin{bmatrix} 2 & 2 & 1 \\ 3 & -2 & -2 \\ -2 & 6 & 3 \end{bmatrix}.$

7.5 $(x,y,z) = (x-y)(1,0,0) + (y-z)(1,1,0) + z(1,1,1)$ and therefore $f(x,y,z) = (z,y,x)$.

$\text{Mat}_B f = \begin{bmatrix} -\frac{4}{3} & -\frac{2}{3} & -1 \\ \frac{2}{3} & \frac{1}{3} & 1 \\ 1 & 1 & 1 \end{bmatrix};$ $\text{Mat}_B g = \begin{bmatrix} \frac{2}{3} & \frac{4}{3} & 2 \\ \frac{2}{3} & \frac{1}{3} & 0 \\ -1 & -1 & -1 \end{bmatrix};$ $\text{Mat}_B fgf = \begin{bmatrix} -\frac{35}{27} & -\frac{40}{27} & -\frac{17}{9} \\ \frac{25}{27} & \frac{17}{27} & \frac{7}{9} \\ 1 & 1 & \frac{4}{3} \end{bmatrix}.$

7.6 A has rank 3 and so is invertible. The matrix of f^{-1} is A^{-1}, namely $\begin{bmatrix} -\frac{9}{2} & -6 & 2 \\ \frac{1}{2} & 1 & 0 \\ \frac{5}{2} & 3 & -1 \end{bmatrix}.$

7.7 Use Theorem 7.3 and induction.   **7.8** $\begin{bmatrix} 1 & 0 & 1 & 0 \\ 0 & 0 & 1 & 1 \\ 0 & 0 & 0 & 1 \\ 1 & 1 & 0 & 0 \end{bmatrix}.$

7.9 $A = \begin{bmatrix} 0 & 1 & 0 \\ -1 & 0 & 0 \\ 0 & 0 & 1 \end{bmatrix};$ $B = \begin{bmatrix} 0 & 0 & -1 \\ -1 & 0 & 0 \\ 1 & 1 & 1 \end{bmatrix}.$ The matrix X represents the identity map relative to a change of reference from the first basis to the second. Since

$$\begin{aligned} (1,0,0) &= \tfrac{1}{2}(1,1,0) - \tfrac{1}{2}(0,1,1) + \tfrac{1}{2}(1,0,1) \\ (0,1,0) &= \tfrac{1}{2}(1,1,0) + \tfrac{1}{2}(0,1,1) - \tfrac{1}{2}(1,0,1) \\ (0,0,1) &= -\tfrac{1}{2}(1,1,0) + \tfrac{1}{2}(0,1,1) + \tfrac{1}{2}(1,0,1) \end{aligned}$$

we see that $X = \frac{1}{2} \begin{bmatrix} 1 & 1 & -1 \\ -1 & 1 & 1 \\ 1 & -1 & 1 \end{bmatrix}.$

7.10 We have to determine an ordered basis $B = \{b_1, b_2, b_3\}$ of $\mathbb{R}^3$ such that P is the transition matrix from B to the natural ordered basis $\{e_1, e_2, e_3\}$. $B = \{e_2 + e_3, e_1 + e_3, e_1 + e_2\}$.

7.11 If A is similar to B then there is an invertible matrix P such that $A = P^{-1}BP$. Since P is invertible, so is P'. Then $A' = P'B'(P^{-1})' = [(P^{-1})']^{-1}B'(P^{-1})'$ and so A' is similar to B'.

7.12 If $A = P^{-1}BP$ then by induction $A^k = P^{-1}B^kP$.

7.13 $\begin{bmatrix} \cos\vartheta & -\sin\vartheta \\ \sin\vartheta & \cos\vartheta \end{bmatrix} = \begin{bmatrix} 1 & i \\ i & 1 \end{bmatrix}^{-1} \begin{bmatrix} e^{i\vartheta} & 0 \\ 0 & e^{-i\vartheta} \end{bmatrix} \begin{bmatrix} 1 & i \\ i & 1 \end{bmatrix}.$

7.14 The respective matrices are

(1) $\begin{bmatrix} 0 & 1 & 0 & 0 & \dots & 0 \\ 0 & 0 & 2 & 0 & \dots & 0 \\ 0 & 0 & 0 & 3 & \dots & 0 \\ \vdots & \vdots & \vdots & \vdots & & \vdots \\ 0 & 0 & 0 & 0 & \dots & n \\ 0 & 0 & 0 & 0 & \dots & 0 \end{bmatrix}$; (2) $\begin{bmatrix} 0 & 0 & 0 & \dots & 0 & 0 \\ n & 0 & 0 & \dots & 0 & 0 \\ 0 & n-1 & 0 & \dots & 0 & 0 \\ 0 & 0 & n-2 & \dots & 0 & 0 \\ \vdots & \vdots & \vdots & & \vdots & \vdots \\ 0 & 0 & 0 & \dots & 1 & 0 \end{bmatrix}$; (3) $\begin{bmatrix} 0 & 1 & -2 & -3 & \dots & -n \\ 0 & 0 & 2 & 0 & \dots & 0 \\ 0 & 0 & 0 & 3 & \dots & 0 \\ \vdots & \vdots & \vdots & \vdots & & \vdots \\ 0 & 0 & 0 & 0 & \dots & n \\ 0 & 0 & 0 & 0 & \dots & 0 \end{bmatrix}.$

7.15 If $x \neq 0$ is such that $f^{p-1}(x) \neq 0$ then for every $k \leqslant p-1$ we have $f^k(x) \neq 0$. To show that $\{x, f(x), \dots, f^{p-1}(x)\}$ is linearly independent, suppose that

$$\lambda_0 x + \lambda_1 f(x) + \dots + \lambda_{p-1} f(x_{p-1}) = 0.$$

Applying f^{p-1} to this we obtain $\lambda_0 f^{p-1}(x) = 0$ whence $\lambda_0 = 0$. Thus we have

$$\lambda_1 f(x) + \dots + \lambda_{p-1} f(x_{p-1}) = 0.$$

Applying f^{p-2} to this we obtain similarly $\lambda_1 = 0$. Continuing in this way we see that every $\lambda_i = 0$ and consequently the set is linearly independent.

If f is nilpotent of index $n = \dim V$ then $\{x, f(x), \dots, f^{n-1}(x)\}$ is a basis of V. The matrix of f relative to this ordered basis is then that in the question. Conversely, if there is an ordered basis with respect to which the matrix of f is of the given form then to see that f is nilpotent of index n it suffices to observe that the matrix M in question is such that $M^n = 0$ and $M^{n-1} \neq 0$.

7.16 We have

$$\begin{aligned} f(1) &= 1 \\ f(X) &= 1 + X \\ f(X^2) &= (1+X)^2 = 1 + 2X + X^2 \\ &\;\;\vdots \\ f(X^n) &= (1+X)^n = 1 + \binom{n}{1}X + \binom{n}{2}X^2 + \dots + X^n \end{aligned}$$

and so the matrix of f is $\begin{bmatrix} 1 & 1 & 1 & 1 & \dots & 1 \\ 0 & 1 & 2 & 3 & \dots & \binom{n}{1} \\ 0 & 0 & 1 & 3 & \dots & \binom{n}{2} \\ \vdots & \vdots & \vdots & \vdots & & \vdots \\ 0 & 0 & 0 & 0 & \dots & 1 \end{bmatrix}.$

7.17 We have

$$f(1) = \begin{bmatrix} 0 & 1 \\ 0 & 0 \end{bmatrix} = -\begin{bmatrix} 1 & 0 \\ 0 & 0 \end{bmatrix} + \begin{bmatrix} 1 & 1 \\ 0 & 0 \end{bmatrix}$$

$$f(X) = \begin{bmatrix} 1 & 0 \\ 1 & 0 \end{bmatrix} = \begin{bmatrix} 1 & 0 \\ 0 & 0 \end{bmatrix} - \begin{bmatrix} 1 & 1 \\ 0 & 0 \end{bmatrix} + \begin{bmatrix} 1 & 1 \\ 1 & 0 \end{bmatrix}$$

$$f(1+X) = \begin{bmatrix} 1 & 1 \\ 0 & 1 \end{bmatrix} = \begin{bmatrix} 1 & 1 \\ 0 & 0 \end{bmatrix} - \begin{bmatrix} 1 & 1 \\ 1 & 0 \end{bmatrix} + \begin{bmatrix} 1 & 1 \\ 1 & 1 \end{bmatrix}$$

and so Mat $f = \begin{bmatrix} -1 & 1 & 0 \\ 1 & -1 & 1 \\ 0 & 1 & -1 \\ 0 & 0 & 1 \end{bmatrix}$. The rank of this matrix is 3 and so dim Ker $f = 0$.

$$\text{Im } f = \text{Span} \left\{ \begin{bmatrix} 0 & 1 \\ 0 & 0 \end{bmatrix}, \begin{bmatrix} 1 & 0 \\ 1 & 0 \end{bmatrix}, \begin{bmatrix} 1 & 0 \\ 0 & 1 \end{bmatrix} \right\}.$$

7.18 $\begin{bmatrix} 1 & 0 & 0 & 0 \\ -1 & 0 & -1 & 1 \\ 0 & 0 & 0 & 0 \\ 0 & 0 & -1 & 1 \end{bmatrix}$; $\begin{bmatrix} 1 & 0 & 0 & 0 \\ 0 & 0 & 0 & 0 \\ 0 & 0 & 1 & 0 \\ 0 & 0 & 0 & 0 \end{bmatrix}$.

7.19 $(1), (2), (3)$ are routine. If A and B are similar then there is an invertible P such that $A = P^{-1}BP$. By (3), $\text{tr}(A) = \text{tr}(P^{-1}BP) = \text{tr}(P^{-1}PB) = \text{tr}(B)$. For the last part, observe for example that $\text{tr}\begin{bmatrix} 1 & 0 \\ 0 & -1 \end{bmatrix} = 0 = \text{tr}\begin{bmatrix} 0 & 0 \\ 0 & 0 \end{bmatrix}$ but these matrices are not similar.

8.1 All expressions are the same.   **8.2** Routine.

8.3 $\begin{pmatrix} 1 & 2 & 3 & 4 & 5 & 6 & 7 & 8 \\ 1 & 4 & 5 & 8 & 7 & 3 & 2 & 6 \end{pmatrix}$; $\begin{pmatrix} 1 & 2 & 3 & 4 & 5 & 6 & 7 & 8 \\ 1 & 2 & 3 & 4 & 5 & 6 & 7 & 8 \end{pmatrix}$.

8.4 Even; odd.   **8.5** -28.   **8.6** $26 + 13\lambda + \lambda^2 - \lambda^3$.   **8.8** 400.

8.9 If the elementary matrix is obtained by interchanging two rows (columns) then the determinant is -1; by multiplying a row (column) by λ it is λ; and by adding λ times one row (column) to another it is 1.

8.10 By induction. For the inductive step, use a Laplace expansion via the first column.

8.11 $x \neq \frac{1}{2}$.

8.12 Use a Laplace expansion via the first row, then via the first column of the resulting 5×5 matrix, then by the first row of the resulting 4×4 matrix, and so on.

8.13 Add the first row to all the others; the answer is $n!$.

8.14 $\det \begin{bmatrix} 0 & 0 & 0 & \dots & 0 & 0 & 1 \\ 0 & 0 & 0 & \dots & 0 & 1 & 0 \\ 0 & 0 & 0 & \dots & 1 & 0 & 0 \\ \vdots & \vdots & \vdots & \dots & \vdots & \vdots & \vdots \\ 0 & 0 & 1 & \dots & 0 & 0 & 0 \\ 0 & 1 & 0 & \dots & 0 & 0 & 0 \\ 1 & 0 & 0 & \dots & 0 & 0 & 0 \end{bmatrix} = (-1)^{(n-1)+(n-2)+\dots+1} = (-1)^{\frac{1}{2}n(n-1)}.$

8.15 For example, take $\rho_3 - 2\rho_2$ then use a Laplace expansion via the first column.

8.16 $P^{-1} = \begin{bmatrix} -2 & 4 & 1 \\ 1 & -2 & 4 \\ 4 & 1 & -2 \end{bmatrix}$ and $P^{-1}AP = \begin{bmatrix} 9a & 0 & 0 \\ 0 & 9b & 0 \\ 0 & 0 & 9c \end{bmatrix}$. $\det A = 3^6 abc.$

8.17 $\det A^p = (\det A)^p = 0$ and so $\det A = 0$.

8.18 $\begin{bmatrix} d & -b \\ -c & a \end{bmatrix}$; $\begin{bmatrix} bc - f^2 & fg - hc & hf - bg \\ fg - hc & ac - g^2 & hg - af \\ hf - bg & gh - af & ab - h^2 \end{bmatrix}$; $\begin{bmatrix} 1 & 0 & -1 \\ 0 & 0 & 0 \\ -1 & 0 & 1 \end{bmatrix}$.

8.19 The first two have determinant 1 so adjugate=inverse; these are respectively

$\begin{bmatrix} 1 & 1 & -3 \\ 0 & 1 & -1 \\ -1 & -2 & 5 \end{bmatrix}$; $\begin{bmatrix} 4 & 1 & -3 \\ 1 & 1 & -1 \\ -11 & -4 & 9 \end{bmatrix}$.

The third has determinant 6 and inverse $\frac{1}{6}\begin{bmatrix} 6 & 0 & 0 \\ -3 & 3 & 0 \\ 0 & -2 & 2 \end{bmatrix}$.

8.21 Since $A \cdot \text{adj } A = (\det A)I_n$ we have

$$(\det A)(\det \text{adj } A) = \det[(\det A)I_n] = (\det A)^n.$$

A being invertible, $\det A \neq 0$ and so $\det \operatorname{adj} A = (\det A)^{n-1}$.

8.22 $AB \operatorname{adj} AB = (\det AB)I_n = (\det A \det B)I_n$ and so $\operatorname{adj} AB = B^{-1}A^{-1} \det A \det B = B^{-1} \operatorname{adj} A \det B = B^{-1} \det B \operatorname{adj} A = \operatorname{adj} B \operatorname{adj} A$.

8.23 $\operatorname{adj}(\operatorname{adj} A) = \det \operatorname{adj} A \cdot (\operatorname{adj} A)^{-1}$. By Exercise 8.21 and the fact that $(\operatorname{adj} A)^{-1} = \dfrac{1}{\det A}A$ we obtain $\operatorname{adj}(\operatorname{adj} A) = (\det A)^{n-2}A$.

8.24 If A is upper triangular then $a_{ij} = 0$ when $i > j$. Thus, for $i < j$ we have $\det A_{ij} = 0$. Consequently, for $i > j$, $[\operatorname{adj} A]_{ij} = (-1)^{i+j} \det A_{ji} = 0$ and so $\operatorname{adj} A$ is also upper triangular.

8.25 For a symmetric matrix, $A_{ij} = A_{ji}$.

8.26 Observe that $[\overline{\operatorname{adj} A}]_{ji} = (-1)^{i+j} \det \overline{A}_{ji} = (-1)^{i+j} \det \overline{A}'_{ij} = [\operatorname{adj} \overline{A}']_{ij}$ and so $\overline{\operatorname{adj} A}' = \operatorname{adj} \overline{A}' = \operatorname{adj} A$.

8.27 $\det A = (x-a)(x-b)(x-c)(a-b)(b-c)(c-a)$.

8.28 $\det = (x-a)^3(x+3a)$ so the solutions are $x = a$ and $x = -3a$.

8.29 Start with $c_1 - yc_2$.

8.30 Begin by adding all the rows to the first row, thereby obtaining a factor $n - 1$. Then, for $i > 1$, take $\rho_i - \rho_1$.

8.31 Begin by adding all the rows to the first row, thereby obtaining a factor $na + b$. Then, for $i > 1$, take $\rho_i - a\rho_1$.

8.32 Begin by subtracting the first column from the others. Then do Laplace expansions via the rows. The solutions are $x = 0, \dots, n-1$.

8.33 For $\det B_n$ first subtract the first row from the other rows, then subtract the last column from the other columns.

For $\det A_n$ first add the last column to the first column. Now do a Laplace expansion via the new first column and use the result for $\det B_{n-1}$.

For the last part, use the previous formula and induction.

8.34 Use a Laplace expansion via the first row of A_{n+2}. The last part follows by induction and basic trigonometry.

8.35 For $\det A_n$ subtract the last column of A_n from the other columns and then use a Laplace expansion via the first row.

For the recurrence formula for $\det B_n$ use a Laplace expansion via the last column.

For the last part use induction.

8.36 We have

$$
\det \begin{bmatrix} A & B \\ B & A \end{bmatrix} = \det \begin{bmatrix} A & B \\ B-A & A-B \end{bmatrix} = \det \begin{bmatrix} A+B & B \\ 0 & A-B \end{bmatrix}
$$
$$
= \det \begin{bmatrix} I & B \\ 0 & A-B \end{bmatrix}\begin{bmatrix} A+B & 0 \\ 0 & I \end{bmatrix}
$$
$$
= \det \begin{bmatrix} I & B \\ 0 & A-B \end{bmatrix} \det \begin{bmatrix} A+B & 0 \\ 0 & I \end{bmatrix}
$$
$$
= \det(A-B)\det(A+B).
$$

8.37 Clearly, $\begin{bmatrix} A & 0 \\ B & C \end{bmatrix}\begin{bmatrix} P & Q \\ R & S \end{bmatrix} = \begin{bmatrix} AP & AQ \\ BP+CR & BQ+CS \end{bmatrix}$ and this is of the given form if and only if

$$
A = P^{-1}, \; BP + CR = 0, \; BQ + CS = S - RP^{-1}Q.
$$

Since $B = -CRP^{-1}$ we have $C(S - RP^{-1}Q) = S - RP^{-1}Q$, so we can choose $C = I$ and then $B = -RP^{-1}$. Now clearly

$$\det N \det M = \det \begin{bmatrix} I & P^{-1}Q \\ 0 & S - RP^{-1}Q \end{bmatrix} = \det \begin{bmatrix} I & 0 \\ 0 & S - RP^{-1}Q \end{bmatrix} = \det(S - RP^{-1}Q)$$

i.e. $\det P^{-1} \det M = \det(S - RP^{-1}Q)$, so

$$\det M = \det P \det(S - RP^{-1}Q) = \det(PS - PRP^{-1}Q).$$

If now $PR = RP$ then $PRP^{-1} = R$ and so $\det M = \det(PS - RQ)$. Likewise, we also have

$$\det M = \det(S - RP^{-1}Q)\det P = \det(SP - RP^{-1}QP)$$

so if $PQ = QP$ then $\det M = \det(SP - RQ)$.

8.38 Follow the hint. 28; -18; $n!$.

9.1 (1) The eigenvalues are $1, 2, 3$ each of algebraic multiplicity 1.

(2) The only eigenvalue is 1, of algebraic multiplicity 3.

(3) The eigenvalues are $2, 1 + i, 2 - 2i$, each of algebraic multiplicity 1.

9.2 Suppose that $Ax = \lambda x$ where $x \neq 0$. If λ were 0 then we would have the contradiction $x = A^{-1}Ax = A^{-1}0 = 0$. Hence $\lambda \neq 0$.

Now $x = A^{-1}Ax = A^{-1}\lambda x = \lambda A^{-1}x$ and so $A^{-1}x = \lambda^{-1}x$; i.e. λ^{-1} is an eigenvalue of A^{-1}.

9.3 If $Ax = \lambda x$ then by induction we have $A^k x = \lambda^k x$. Thus, for any polynomial $p(X) = a_0 + a_1 X + \cdots + a_n X^n$ we have $p(A)x = p(\lambda)x$ whence $p(\lambda)$ is an eigenvalue of $p(A)$.

9.4 Proceed as in Example 9.3.

(1) The eigenvalues are $0, 1, 2$, each of algebraic multiplicity 1.

$$E_0 = \text{Span}\left\{ \begin{bmatrix} x \\ 0 \\ -x \end{bmatrix} ; x \neq 0 \right\} \text{ so a basis is } \left\{ \begin{bmatrix} 1 \\ 0 \\ -1 \end{bmatrix} \right\}.$$

$$E_1 = \text{Span}\left\{ \begin{bmatrix} 0 \\ x \\ 0 \end{bmatrix} ; x \neq 0 \right\} \text{ so a basis is } \left\{ \begin{bmatrix} 0 \\ 1 \\ 0 \end{bmatrix} \right\}.$$

$$E_2 = \text{Span}\left\{ \begin{bmatrix} x \\ 0 \\ x \end{bmatrix} ; x \neq 0 \right\} \text{ so a basis is } \left\{ \begin{bmatrix} 1 \\ 0 \\ 1 \end{bmatrix} \right\}.$$

(2) The eigenvalues are $0, 1, -1$, each of algebraic multiplicity 1.

$$E_0 = \text{Span}\left\{ \begin{bmatrix} x \\ -x \\ x \end{bmatrix} ; x \neq 0 \right\} \text{ so a basis is } \left\{ \begin{bmatrix} 1 \\ -1 \\ 1 \end{bmatrix} \right\}.$$

$$E_1 = \text{Span}\left\{ \begin{bmatrix} -x \\ -2x \\ x \end{bmatrix} ; x \neq 0 \right\} \text{ so a basis is } \left\{ \begin{bmatrix} -1 \\ -2 \\ 1 \end{bmatrix} \right\}.$$

$$E_{-1} = \text{Span}\left\{ \begin{bmatrix} 2x \\ -x \\ x \end{bmatrix} ; x \neq 0 \right\} \text{ so a basis is } \left\{ \begin{bmatrix} 2 \\ -1 \\ 1 \end{bmatrix} \right\}.$$

9.5 The eigenvalues are 1 and 2, of algebraic multiplicities 2, 1 respectively.

$$E_1 \text{ has basis } \left\{ \begin{bmatrix} 1 \\ 0 \\ -1 \end{bmatrix} , \begin{bmatrix} 0 \\ 1 \\ -1 \end{bmatrix} \right\} \text{ and } E_2 \text{ has basis } \left\{ \begin{bmatrix} 3 \\ 1 \\ -5 \end{bmatrix} \right\}.$$

9.6 (1) Relative to the natural ordered basis of $\mathbb{R}^3$ the matrix of f is $\begin{bmatrix} 1 & 2 & 2 \\ 0 & 2 & 1 \\ -1 & 2 & 2 \end{bmatrix}$. The eigen-

values are $1, 2$ of algebraic multiplicities $1, 2$.

(2) Relative to the natural ordered basis of $\mathbb{R}^3$ the matrix of f is $\begin{bmatrix} 0 & 1 & 1 \\ 0 & 0 & 0 \\ 1 & 1 & 0 \end{bmatrix}$. The eigenvalues

are $0, 1, -1$ each of algebraic multiplicity 1.

9.7 The first two matrices are those of Exercise 9.4. Suitable matrices P are therefore, respec-

tively, $\begin{bmatrix} 1 & 0 & 1 \\ 0 & 1 & 0 \\ -1 & 0 & 1 \end{bmatrix}$ and $\begin{bmatrix} 1 & -1 & 2 \\ -1 & -2 & -1 \\ 1 & 1 & 1 \end{bmatrix}$.

Proceeding similarly with the other two, we see that suitable matrices P are, respectively,

$\begin{bmatrix} 3 & 1 & 2 \\ 1 & 2 & 1 \\ 1 & 1 & 1 \end{bmatrix}$ and $\begin{bmatrix} 1 & 1 & -3 \\ 0 & 1 & -1 \\ -1 & -2 & 5 \end{bmatrix}$.

9.8 Let $\det A_n = a_n$. Use a Laplace expansion via the first column to obtain the recurrence

relation $a_n = a_{n-1} + 20a_{n-2}$. Consider therefore the system of difference equations

$$a_n = a_{n-1} + 20b_{n-1}$$
$$b_n = a_{n-1}.$$

The matrix of the system is $A = \begin{bmatrix} 1 & 20 \\ 1 & 0 \end{bmatrix}$ which has two distinct eigenvalues, -4 and 5. By

Theorem 9.8, the matrix $P = \begin{bmatrix} 20 & 20 \\ -5 & 4 \end{bmatrix}$ is invertible and such that $P^{-1}AP = \begin{bmatrix} -4 & 0 \\ 0 & 5 \end{bmatrix}$. Now

$P^{-1} = \dfrac{1}{180} \begin{bmatrix} 4 & -20 \\ 5 & 20 \end{bmatrix}$ and so

$$A^n = P \begin{bmatrix} (-4)^n & 0 \\ 0 & 5^n \end{bmatrix} P^{-1} = \frac{1}{9} \begin{bmatrix} 5^{n+1} - (-4)^{n+1} & 4 \cdot 5^{n+1} + 5 \cdot (-4)^{n+1} \\ 4 \cdot 5^n + 5 \cdot (-4)^n & 4 \cdot 5^n + 5 \cdot (-4)^n \end{bmatrix}.$$

Finally, $\begin{bmatrix} a_n \\ b_n \end{bmatrix} = A^{n-2} \begin{bmatrix} a_2 \\ b_2 \end{bmatrix} = A^{n-2} \begin{bmatrix} 21 \\ 1 \end{bmatrix} = \frac{1}{9} \begin{bmatrix} 5^{n+1} - (-4)^{n+1} \\ 5^n - (-4)^n \end{bmatrix}$ from which the result follows.

9.9 $(1 - X)^3$; $-X^3 + X^2 + 3X - 1$; $-X^3 - 3X^2 - 6X - 34$.

9.10 We have

$$\begin{aligned}
(I_n - BA)\big[I_n + B(I_n - AB)^{-1}A\big] &= I_n - BA + (I_n - BA)B(I_n - AB)^{-1}A \\
&= I_n - BA + (B - BAB)(I_n - AB)^{-1}A \\
&= I_n - BA + B(I_n - AB)(I_n - AB)^{-1}A \\
&= I_n - BA + BA \\
&= I_n.
\end{aligned}$$

Hence $(I_n - BA)^{-1} = I_n + B(I_n - AB)^{-1}A$.

The last part follows from the fact that λ is an eigenvalue of XY if and only if $XY - \lambda I_n$ is

not invertible which, in the case where $\lambda \neq 0$, is equivalent to $I_n - \lambda^{-1}XY$ being not invertible.

9.11 $P^{-1}AP = \begin{bmatrix} e^{i\vartheta} & 0 \\ 0 & e^{-i\vartheta} \end{bmatrix}$.   **9.12** $P = \begin{bmatrix} 1 & -1 & 1 \\ -\sqrt{2} & 0 & \sqrt{2} \\ 1 & 1 & 1 \end{bmatrix}$; $P = \begin{bmatrix} 0 & 1 & -1 \\ 0 & 1 & 1 \\ 1 & 0 & 0 \end{bmatrix}$.

9.13 Proceed as in Example 9.10. By Theorem 9.8 there is an invertible P such that $P^{-1}AP = \begin{bmatrix} \lambda_1 & 0 \\ 0 & \lambda_2 \end{bmatrix}$ and, since $\lambda_1 + \lambda_2 = \alpha$, we have $P = \begin{bmatrix} \beta & \beta \\ -\lambda_2 & -\lambda_1 \end{bmatrix}$. When the eigenvalues are not distinct, use the expression for A^n given on page 164.

9.14 $\dfrac{1}{6} \begin{bmatrix} 4 + 2 \cdot 4^n & 3 \cdot 4^n & -4 + 4^n \\ -2 + 2 \cdot 4^n & 3 \cdot 4^n & 2 + 4^n \\ -2 + 2 \cdot 4^n & 3 \cdot 4^n & 2 + 4^n \end{bmatrix}$.   **9.15** $\begin{bmatrix} x_n \\ y_n \end{bmatrix} = \begin{bmatrix} \frac{1}{13}[6(-7)^{n-1} - 6^n] \\ -\frac{1}{13}[9(-7)^{n-1} + 4 \cdot 6^{n-1}] \end{bmatrix}$.

9.16 Diagonalising A in the usual way we have $P^{-1}AP = D$ where

$$P = \begin{bmatrix} 1 & 1 & 0 \\ 1 & -1 & 1 \\ 0 & 0 & 1 \end{bmatrix}, \quad P^{-1} = \begin{bmatrix} \frac{1}{2} & \frac{1}{2} & -1 \\ \frac{1}{2} & -\frac{1}{2} & 0 \\ 0 & 0 & 1 \end{bmatrix}, \quad D = \begin{bmatrix} 4 & 0 & 0 \\ 0 & 9 & 0 \\ 0 & 0 & 4 \end{bmatrix}.$$

Now the matrix $E = \begin{bmatrix} 2 & 0 & 0 \\ 0 & 3 & 0 \\ 0 & 0 & 2 \end{bmatrix}$ is such that $E^2 = D$ and consequently the matrix $B = PEP^{-1}$

has the property $B^2 = A$. A simple computation shows that $B = \begin{bmatrix} \frac{5}{2} & -\frac{1}{2} & \frac{1}{2} \\ -\frac{1}{2} & \frac{5}{2} & -\frac{1}{2} \\ 0 & 0 & 2 \end{bmatrix}$.

9.17 For the first part, use induction. The result clearly holds for $r = 1$. Suppose that it holds for r. Then

$$|[A^{r+1}]_{ij}| = \left| \sum_{t=1}^{n} a_{it}[A^r]_{tj} \right| \leqslant \sum_{t=1}^{n} |a_{it}||[A^r]_{tj}| \leqslant \sum_{t=1}^{n} k \cdot k^r n^{r-1} = k^{r+1} n^{r-1} \sum_{t=1}^{n} 1 = k^{r+1} n^r.$$

(1) If $|\beta| < \dfrac{1}{nk}$ then

$$1 + |\beta||[A_{ij}]| + |\beta^2||[A^2]_{ij}| + \cdots + |\beta^r||[A^r]_{ij}| + \cdots$$
$$\leqslant 1 + k|\beta| + k^2 n|\beta|^2 + \cdots + k^r n^{r-1}|\beta|^r + \cdots$$
$$= 1 + k|\beta|(1 + kn|\beta| + \cdots + k^{r-1} n^{r-1}|\beta|^{r-1} + \cdots)$$

which is less than or equal to a geometric series which converges. Thus we see that if $|\beta| < \dfrac{1}{nk}$ the $S_\beta(A)$ is absolutely convergent, hence convergent.

(2) If $S_\beta(A)$ is convergent then $\lim_{t \to \infty} \beta^t A^t = 0$ so

$$(I_n - \beta A)(I_n + \beta A + \beta^2 A^2 + \cdots) = \lim_{t \to \infty}(I_n - \beta A)(I_n + \beta A + \cdots + b^n A^n)$$
$$= \lim_{t \to \infty}(I_n - \beta^{t+1} A^{t+1})$$
$$= I_n.$$

Consequently $I_n - \beta A$ has an inverse which is the sum of the series.

For the last part, let λ be an eigenvalue of A. Then $\lambda I_n - A$ is not invertible. Suppose, by way of obtaining a contradiction, that $|\lambda| > nk$. Then $\dfrac{1}{|\lambda|} < \dfrac{1}{nk}$. Consequently, if we let $\beta = \lambda^{-1}$ we have, by (1), that $S_\beta(A)$ converges and so, by (2), $I_n - \beta A = I_n - \lambda^{-1}A$ is invertible. It follows that $\lambda I_n - A$ is invertible, a contradiction. Hence we must have $|\lambda| \leqslant nk$.

10.1 $m_A(X) = a_0 + a_1 X + \cdots + a_p X^p$. We must have $a_0 \neq 0$ since otherwise $0 = m_A(A) = a_1 A + \cdots + A_p A^p$ and therefore, since A is invertible, $a_1 I_n + a_2 A + \cdots + a_p A^{p-1} = 0$ which contradicts the fact that $m_A(X)$ is the minimum polynomial of A.

Now $0 = a_0 I_n + a_1 A + \cdots + a_p A^p$ can be written $A(a_1 + a_2 A + \cdots + a_p A^{p-1}) = -a_0 I_n$ whence, since A is invertible, we have $A^{-1} = -\dfrac{1}{a_0}(a_1 + a_2 A + \cdots + a_p A^{p-1})$. Thus A^{-1} is a linear combination of $I_n, A, \ldots, A^{p-1}$.

10.2 For each of the matrices the characteristic and minimum polynomials coincide; they are respectively $(X-1)^3$; $-X^3 + X^2 + X - 3$; $-X^3 + X^2 - X + 2$.

10.3 $\chi_A(X) = \det[A - XI_n] = a_0 + a_1 X + \cdots + A_n X^n$. Taking $X = 0$ we obtain $\det A = a_0$.

10.4 We have

$$\chi_{R_\vartheta}(X) = X^2 - 2X\cos\vartheta + 1 = (X - \cos\vartheta - i\sin\vartheta)(X - \cos\vartheta + i\sin\vartheta).$$

Then $\chi_{R_\vartheta}(X) = m_{R_\vartheta}(X)$. If ϑ is not an integer multiple of π then $i\sin\vartheta \neq 0$ and so r_ϑ has no real eigenvalues.

10.5 $\chi(D) = (-1)^{n-1}X^{n-1} = m(D)$.

10.6 The matrix of f relative to the natural ordered basis $\{(1,0),(0,1)\}$ is $A = \begin{bmatrix} 1 & 4 \\ -\frac{1}{2} & -1 \end{bmatrix}$.

$m_f(X) = m_A(X) = X^2 - 3$.

10.7 Consider the matrix $\begin{bmatrix} 1 & 0 & 1 \\ 0 & 2 & 1 \\ -1 & 0 & 3 \end{bmatrix}$. The only eigenvalue is 2 (of algebraic multiplicity 3).

The geometric multiplicity of this eigenvalue is 2 so f is not diagonalisable.

10.8 For all positive integers i we have $A^{i+1} = (PQ)^{i+1} = P(QP)^i Q = PB^i Q$. If now $h(X) = z_0 + z_1 X + \cdots + z_n X^n$ then we have

$$\begin{aligned} Ah(A) &= z_0 A + z_1 A^2 + \cdots + z_n A^{n+1} \\ &= z_0 PQ + z_1 PBQ + \cdots + z_n PB^n Q \\ &= P(z_0 I_n + z_1 B + \cdots + z_n B^n)Q \\ &= Ph(B)Q. \end{aligned}$$

It follows immediately that $Am_B(A) = Pm_B(B)Q = 0$. Similarly, we have $Bm_A(B) = 0$. Consequently, $m_A(X)|Xm_B(X)$ and $m_B(X)|Xm_A(X)$, and so we can write $Xm_B(X) = p(X)m_A(X)$ and $Xm_A(X) = q(X)m_B(X)$. Comparing the degrees of each of these equations, we deduce that $\deg p + \deg q = 2$. Thus, either $\deg p = 0$ in which case $p(X) = 1$ and $Xm_B(X) = m_A(X)$, or $\deg q = 0$ in which case $q(X) = 1$ and $Xm_A(X) = m_B(X)$, or $\deg p = \deg q = 1$ in which case $m_A(X) = m_B(X)$.

10.9 The matrix can be written as a product PQ where P is the column matrix $\begin{bmatrix} 1 & 2 & \cdots & r \end{bmatrix}'$ and Q is the row matrix $\begin{bmatrix} 1 & 1 & \cdots & 1 \end{bmatrix}$. Note that then $B = QP$ is the 1×1 matrix whose entry is $\frac{1}{2}r(r+1)$. We have $m_B(X) = -\frac{1}{2}r(r+1) + X$. Clearly, $m_A(X) \neq m_B(X)$ and $m_B(X) \neq Xm_A(X)$. Thus, by the previous exercise, we must have $m_A(X) = Xm_B(X) = -\frac{1}{2}X + X^2$.

10.10 The matrix of f relative to the ordered basis $\{1, 1+X, 1+X-X^2\}$ is

$$A = \begin{bmatrix} -1 & 0 & 0 \\ 2 & 5 & 6 \\ -2 & -3 & -4 \end{bmatrix}.$$

$\chi_A(X) = (X-2)(X+1)^2 = m_A(X)$.

Index